Molecular Biology Techniques:
A Classroom Laboratory Manual

Molecular Biology Techniques: A Classroom Laboratory Manual
THIRD EDITION

Susan Carson
Heather B. Miller
D. Scott Witherow

AMSTERDAM • BOSTON • HEIDELBERG • LONDON • NEW YORK • OXFORD
PARIS • SAN DIEGO • SAN FRANCISCO • SINGAPORE • SYDNEY • TOKYO
Academic Press is an imprint of Elsevier

Academic Press is an imprint of Elsevier
32 Jamestown Road, London NW1 7BY, UK
225 Wyman Street, Waltham, MA 02451, USA
525 B Street, Suite 1800, San Diego, CA 92101-4495, USA

First edition 1997
Second edition 2006
Third edition 2012

Notice
No responsibility is assumed by the publisher for any injury and/or damage to persons or property as a matter of
products liability, negligence or otherwise, or from any use or operation of any methods, products, instructions or
ideas contained in the material herein. Because of rapid advances in the medical sciences, in particular, independent
verification of diagnoses and drug dosages should be made

British Library Cataloguing-in-Publication Data
A catalogue record for this book is available from the British Library

Library of Congress Cataloging-in-Publication Data
A catalog record for this book is available from the Library of Congress

ISBN: 978-0-12-385544-2

For information on all Academic Press publications
visit our website at elsevierdirect.com

Typeset by MPS Limited, a Macmillan Company, Chennai, India
www.macmillansolutions.com

Printed and bound by CPI Group (UK) Ltd, Croydon, CR0 4YY

Transferred to Digital Print 2012

Working together to grow
libraries in developing countries

www.elsevier.com | www.bookaid.org | www.sabre.org

ELSEVIER   BOOK AID
            International   Sabre Foundation

Contents

Preface

Recombinant DNA technology touches many aspects of our lives. From drought-tolerant crops to biofuels to pharmaceuticals, genetically modified organisms surround us. Throughout its history, biotechnology has relied heavily on DNA cloning to produce protein products more efficiently or to make modified genes or combinations of genes. *E. coli* or other hosts serve as a cellular "factories" to churn out large amounts of protein. Human insulin, for example, is produced recombinantly using many of the same techniques described in this book.

In the seven years since the most recent edition of *Manipulation and Expression of Recombinant DNA: A Laboratory Manual* was published, molecular biology research and its tools have greatly advanced. Genomic sequences of thousands of organisms are now known. Cloning by PCR has all but replaced traditional cloning methods. DNA sequencing is cheap and accessible. Tools to study that elusive nucleic acid, RNA, are now established as mainstream molecular biology techniques.

Emerging evidence suggests that the central dogma of molecular biology, "DNA makes RNA makes protein," is more complex and has more exceptions than we ever imagined. The transition of a human gene from DNA to RNA may not be as faithful as we once thought. Recently, more research has focused on RNA, and this has led to groundbreaking findings. These discoveries draw an even more complex picture of gene expression. Not only can this multifaceted nucleic acid serve as a "messenger," but it can silence genes, perform catalysis, and generate protein diversity!

This new classroom laboratory manual uses the semester-long project concept introduced in *Manipulation and Expression of Recombinant DNA: A Laboratory Manual*, but the laboratory sessions are updated to reflect current cloning techniques. We also added a new section (five laboratory sessions) on measuring mRNA levels to underscore our conviction that today's molecular biology students need to be confident working with RNA. The laboratory sessions in this new manual *Molecular Biology Techniques: A Classroom Laboratory Manual* reflect the nature of modern science; that is, scientists need the experience to work with macromolecules at each level of gene expression: DNA, RNA and protein.

What will we learn in the next seven years? Students taking this course right now will be at the leading edge of exciting new discoveries. We can't wait to find out!

About the Authors

Left to right: Sue Carson, Heather Miller, Scott Witherow

Dr. Sue Carson is the Academic Coordinator of the Biotechnology Program, Teaching Associate Professor of Plant Biology, and the Director of the NSF-funded Synthetic Biology Research Experience for Undergraduates at North Carolina State University. She graduated from Rutgers University (New Brunswick, NJ) with a B.S. in Biotechnology, and from The University of North Carolina (Chapel Hill, NC) with a Ph.D. in Microbiology. Her area of scientific expertise is in molecular mechanisms of bacterial pathogenesis. Dr. Carson's current work focuses on college-level science education. She has received multiple awards for teaching excellence and innovation and is a member of the Howard Hughes Medical Institute Science Education Alliance, promoting inquiry-guided learning in the college classroom laboratory. She co-authored the molecular biology lab manual *Manipulation and Expression of Recombinant DNA: A Laboratory Manual 2e* (Academic Press, 2006), and has published numerous peer-reviewed papers in the area of course and curriculum development.

Dr. Heather Miller is a teaching postdoctoral fellow in the Biotechnology Program at North Carolina State University. She graduated from Clarion University of Pennsylvania (Clarion, PA) with a B.S. in Molecular Biology/Biotechnology, and from Duke University (Durham, NC) with a Ph.D. in Molecular Genetics and Microbiology. Her area of scientific expertise is RNA biology. Her research has focused on HIV-1 gene expression and the coupling of transcription and splicing in mammalian systems. She has developed and taught multiple biotechnology courses and is engaged in the scholarship of teaching and learning.

Dr. D. Scott Witherow is an Assistant Professor of Chemistry and Biochemistry at the University of Tampa. He graduated from Rollins College (Winter Park, FL) with an A.B. in Chemistry, and from the University of Miami (Miami, FL) with a Ph.D. in Molecular and Cellular Pharmacology. His research has focused primarily on G protein-mediated signal transduction processes in mammalian systems. Following two research postdoctoral fellowships, Dr. Witherow served as a teaching postdoctoral fellow at North Carolina State University, where he published and presented multiple papers in the field of science education and developed a passion for teaching.

Acknowledgements

We would like to thank the many people who contributed to this manual. Melissa Cox was instrumental in piloting the experiments and in critically editing the prep lists. We also thank the North Carolina State University Colleges of Agriculture and Life Sciences, Engineering, Veterinary Medicine, Physical and Mathematical Sciences and Natural Resources, and the Office of the Provost. Their support for the Biotechnology Program led to the development of this manual.

xv

Note to Instructors

These laboratory exercises were developed in the context of the curriculum offered by the North Carolina State University Biotechnology Program (http://biotech.ncsu.edu). Students take our biotechnology core course "Manipulation and Expression of Recombinant DNA" as a prerequisite to more specialized laboratory courses including Real-Time PCR, RNA Interference and Model Organisms, mRNA Transcription and Processing, Protein Purification, Animal Cell Culture, Microarray Technology, Plant Genetic Engineering, Protein–Protein Interactions, Deep Sequencing and others. The laboratories in the core course prepare students well for these specialized courses, and for independent research in a molecular biology laboratory, both at the undergraduate and graduate student levels.

We use the first 14 lab sessions of this manual in our core course, and then a subset of students go on to pursue advanced courses that focus on working with RNA. We chose to include a section on analyzing mRNA levels in this manual. Programs may choose to substitute some of the earlier screening labs with an RNA lab at the end, or add the RNA modules into another course.

Our biotechnology core course is a four-credit lecture/lab course. We meet for two hours once per week for a lecture, and allot four hours for one lab period per week. We recommend this schedule because, for several of the labs, students must inoculate cultures or perform other short activities prior to their lab day. It works out well for them to do so at the end of their lecture period (these activities are referred to as "*interim laboratory sessions*"). The majority of the laboratories do not require the full four hours, but a few of them do (notably Lab Session 6). Additionally, there are a few labs for which incubation times are simply too long to reasonably include in the exercises. In these cases, the steps are included in the protocols with a note that the instructor will perform that particular part of the experiment for the class (for example, the induction of the fusion protein using IPTG, necessary for several laboratories, takes two to four hours). For this reason, we recommend offering the laboratory as an afternoon course so that the instructor can begin the incubations in the morning, rather than the middle of the night.

We use commercially prepared competent cells in our course. If your budget does not permit this, it is possible to prepare "home-made" competent cells and store them in a $-80°C$ freezer. We have included the protocol for preparation of competent cells as Appendix 3.

In this manual, we often refer to "lab stations." This course was designed for students to work in pairs. Each pair of students is assigned to a numbered lab station. Students label all of their experiments, cultures, etc., with their station number, rather than initials.

All antibodies described in this manual are available commercially. The pBIT and pEGFP-N1 plasmids are available at no cost (other than shipping) to institutions of higher education for educational purposes from Dr. Sue Carson at NCSU. Contact Dr. Carson at bit_minor@ncsu.edu, and include in the subject heading "pBIT request." pET-41a is available commercially, and we are not licensed to distribute it.

Instrumentation

Certain lab sessions provide detailed instructions for using a particular brand-name apparatus. Similar equipment from other vendors can be easily substituted. Appropriate instrument-specific instructions should be substituted to minimize student confusion. This is especially true of the DNA agarose gel electrophoresis units, the protein polyacrylamide gel electrophoresis units, the transfer apparatus for western blotting and real-time-capable thermal cycler.

We highly recommend the use of a Nanodrop spectrophotometer. Only 1–2 µl of sample is needed to quantify DNA and RNA. Standard spectrophotometers that use traditional quartz cuvettes require the use of much greater quantities of sample, and in some cases, the sample cannot be diluted for readings because the concentration would be too low. If you do not have access to a Nanodrop spectrophotometer, in some cases where sample amount is limiting, running a small amount of sample on a gel and estimating the quantity is the best method.

Nomenclature

In the literature, the nomenclature for the abbreviations of the enhanced green fluorescent protein gene and its gene product has been inconsistent, at best, and downright confusing at worst. In this publication, we will use "*egfp*" to refer to the gene (either DNA or mRNA) and "EGFP" to refer to the gene product. Likewise, we will use "*gst*" for the glutathione-S-transferase gene and "GST" for its gene product. Bacterial genes discussed in this book will use standard bacterial nomenclature with the gene name lower-case italicized, and the gene product with a capitalized first letter and not italicized. For example the gene for the *lac* repressor is "*lacI*" and its gene product is "LacI."

Introduction

Conceptual Outline for Experiments

Goal: Make a fusion protein by joining genes from two organisms: one from *Escherichia coli* (*gst*) and an enhanced gene derived from the green fluorescent jellyfish *Aequoria victoria* (*egfp*). Expression of the fused gene will produce a single protein in bacteria. The *E. coli* part of the fusion protein will be used as a tag to purify the fusion protein. The *A. victoria* portion of the fusion protein can then be visualized by fluorescence.

Experimental Procedures

Part I: Manipulation of DNA

(See Figure 1 for a diagrammatic representation.)

- Isolate plasmid DNA using cultures of bacteria containing the *E. coli* expression vector pET-41a.
- Use restriction enzymes to cut the pET-41a vector.
- Use PCR to amplify the insert (*A. victoria egfp* DNA) from pEGFP-N1, and restriction digest to form sticky ends.
- Use DNA ligase to "paste" the vector and insert DNA together.
- Introduce the ligated DNA into *E. coli*.

Part II: Screening Transformants

- Identify bacterial transformants that correctly express the *A. victoria* DNA in a fusion protein by probing with a monoclonal antibody.
- Confirm positive clones by polymerase chain reaction.
- Isolate DNA from transformants and digest with restriction enzymes to further validate the presence of the *A. victoria* gene.
- Final confirmation of *egfp*-positive clones by fluorescence.
- Verify that single nucleotide errors did not occur in the *egfp* gene during PCR cloning by DNA sequencing.

Part III: Expression, Detection, and Purification of Recombinant Proteins from Bacteria

- Use sodium dodecyl sulfate-polyacrylamide gel electrophoresis (SDS-PAGE) and western blot analysis to confirm the expression of the fusion protein.

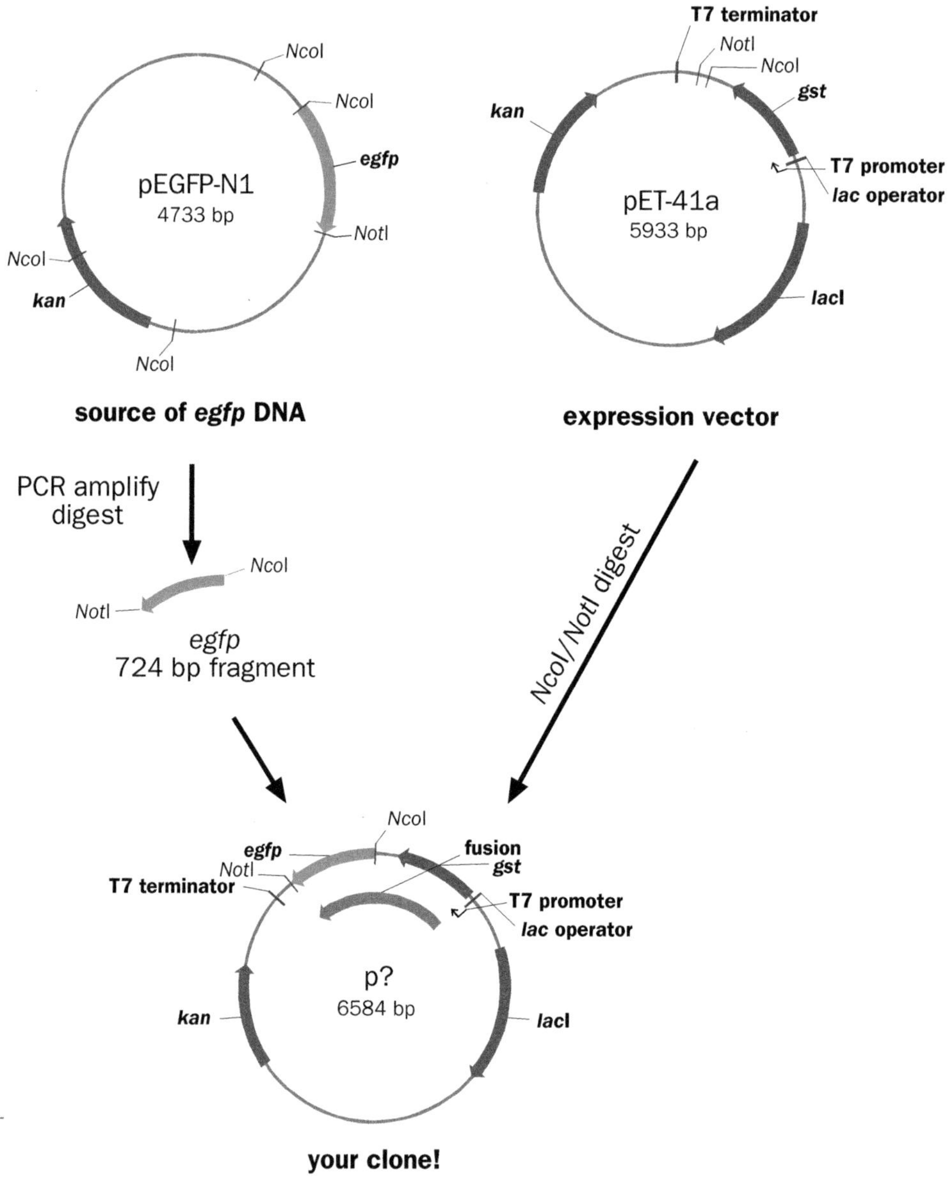

pEGFP-N1
4733 bp

NcoI
NcoI
egfp
NotI
NcoI
kan
NcoI

source of *egfp* DNA

T7 terminator
NotI
NcoI
gst
kan
pET-41a
5933 bp
T7 promoter
lac operator
lacI

expression vector

PCR amplify
digest

NcoI
NotI
egfp
724 bp fragment

NcoI/NotI digest

NcoI
egfp
NotI
fusion
gst
T7 terminator
T7 promoter
lac operator
p?
6584 bp
kan
lacI

your clone!

FIG. 1
Cloning procedure diagram

- Induce a large-scale culture of the transformed bacteria with isopropyl-β-D-thiogalactopyranoside (IPTG) to make large amounts of the fusion protein.
- Purify the fusion protein on a substrate affinity column.
- Perform protein quantification of eluted fractions.
- Use SDS-PAGE of purification fractions to check for purity and degradation.

Part IV: Analysis of mRNA Levels

- Purify total RNA from a positive bacterial clone.
- Perform reverse transcription polymerase chain reaction to quantify mRNA levels after induction with IPTG and/or lactose.

Laboratory Safety

Hazards that you may be exposed to during the course of the laboratory exercises include working with toxic compounds and ultraviolet (UV) irradiation. Special precautions must be taken when working with recombinant DNA. To ensure the safety and well-being of students and support staff, the following rules will be strictly enforced.

A reckless attitude about the use of equipment or the safety of others will cause you to be dropped from the course.

The following rules *must* be observed at all times in the laboratory:

1. No drinking or eating is allowed in the laboratory. No food or drinks should be brought into the laboratory.
2. Safety glasses provided must be worn at all times in the laboratory. No sandals or other open-toed shoes are allowed in the laboratory at any time.
3. Always wear gloves when working with any hazardous or potentially hazardous substance. Remove your gloves before leaving the laboratory and change your gloves frequently to prevent contamination of equipment, etc., with caustic agents.
4. Long hair must be tied back at all times, and avoid loose-fitting clothing to avoid hazards associated with open flames, sterile cultures and hazardous chemicals.
5. Dispose of microorganisms, including the tubes used for their growth, in bags marked "BIOHAZARD" for autoclaving. Liquid medium containing microorganisms will be collected in specially marked containers containing bleach.
6. Dispose of glass *only* in properly marked boxes designated for glass disposal. Do not put glass or any sharp object in the autoclavable bags marked biohazard!
7. Keep your lab bench free of unnecessary clutter. Use cabinets and drawers for storing personal items and extra supplies. At the end of the day, your bench should be clean and equipment put away.
8. Wear ear protection when working with the sonicator.
9. Always wear a UV-protective full-face shield when using the transilluminator if it is not in a safety cabinet. Your safety glasses are *not* UV protective. Do not try to analyze your gel on the transilluminator. Observe the gel on the monitor if you are using a digital imager, or take a photograph and analyze the photograph if you do not have a digital imager.
10. Wash your hands thoroughly before you leave the laboratory.
11. All spills should be cleaned up immediately. Notify an instructor if you spill a potentially hazardous chemical or liquid containing live microorganisms.
12. Immediately report all accidents such as spills, cuts, burns, or other injuries to an instructor.
13. Know the location of the fire extinguisher, eye wash station, emergency shower and emergency exits.
14. If you have trouble with a power supply or the leads to a gel, report it to an instructor. If you see someone receiving an electrical shock, use a

non-conducting object, such as a plastic beaker, to break the circuit or you may also receive a shock.

15. Leave all laboratory facilities and equipment in good condition at the end of the class. Before leaving the laboratory, check to make sure that all electrical equipment is turned off and that the gas to the Bunsen burner is turned off.

16. No pets are allowed in the laboratory.

17. Dispose of hazardous chemicals only in designated containers. Do not pour them down the sink.

18. An up-to-date immunization against tetanus is strongly recommended.

19. For a tutorial on general laboratory safety, visit the website http://www.ncsu.edu/project/ungradreshhmi/evaluationModule/login.php (login is free).

General Operating Procedures

- Reagents: Aliquoting reagents and supplies for everyone in the course at the same time is difficult. Your patience and cooperation are greatly appreciated. If you run out of a reagent or enzyme and cannot find it on a lab cart, please ask a technician or TA for more.

- Enzymes: Enzymes are very expensive and can be ruined by prolonged exposure to room temperature or by contamination. Always use a fresh, sterile pipette tip for each enzyme. Keep the enzyme on ice; never put it in a microcentrifuge rack at room temperature. Always add enzymes to your reactions as the last component; addition of enzyme to unbuffered solutions can kill its activity.

- Pipette tips: You will be provided with three boxes of pipette tips: P10, P20/P200 and P1000. When you run out of tips, notify your instructor.

- Repairs: A list of supplies and equipment at your station will be made available to each pair of students at the first lab session. You are responsible for the equipment at your station. If you encounter any difficulties with operation, please ask for assistance. If equipment is damaged in any way, please report it so that we can have it repaired. Barring willful destructiveness, you will not have to pay for repairs.

- Equipment: Please make sure that you understand the proper use of this equipment before attempting to operate it. If in doubt, ask an instructor, not another student. In addition to costly repairs, improper use of this equipment can be very dangerous.

Remember: A reckless attitude to the use of equipment or the safety of others will cause you to be dropped from the course.

PART 1

Manipulation of DNA

The goal of these laboratory exercises is to fuse a jellyfish gene with a bacterial gene and to express a single protein from this hybrid DNA sequence. Why would you want to do this? Molecular shuffling of genetic sequences, or gene cloning, is a powerful tool for understanding biological processes and for biotechnological applications. Using basic tools developed in *Escherichia coli*, we can ask questions about other, more complicated organisms.

Scientists have exploited *E. coli* both as a workhorse for producing DNA and as a source of well-characterized sequences to direct transcription and translation of foreign DNA into protein. With genetic sequence information being produced at a breathtaking rate, the limiting factor is not in sequencing DNA, but in our understanding of the function of the products of these sequences.

In terms of practical biotechnology applications, it can be a huge advantage to clone the gene encoding a difficult-to-purify protein into *E. coli* so that the purification process can be accomplished less expensively and to a greater degree of purity (and oftentimes more ethically, especially if a human gene is involved!). The first recombinant protein to be produced and marketed was human insulin in the early 1980s, which has been invaluable to countless diabetics. The basic tools you will learn in this class will enable you to clone, express and purify recombinant proteins. They will enable you to begin to probe the function of any protein for which a gene has been identified, and will give you the conceptual background needed for tackling more advanced techniques.

Other hosts are now commonly used for cloning DNA and expressing recombinant proteins, such as members of the bacterial genus *Bacillus*, as well as eukaryotic hosts including numerous species of yeast and other fungi, plants, insect cell culture, mammalian cell culture and even whole, live mammals ("pharming"). Many of the recombinant DNA methods used in this course are applicable to cloning in other hosts.

The gene we will be cloning and expressing is the enhanced green fluorescent protein gene, *egfp*. *egfp* is a brightness-enhanced variant of the green fluorescent protein from the jellyfish *Aequoria victoria*.[1] The gene encoding the green fluorescent protein (and its variants, including *egfp*) is widely used as a "reporter gene" or "marker." A reporter gene is a gene that is used to track protein expression. It must have phenotypic expression that is easy to monitor and can be used to study promoter activity or protein localization in different environmental conditions, different tissues, or different developmental stages. Recombinant DNA constructs are made in which the reporter gene is fused to a promoter region of interest and the construct is transformed or transfected into a host cell or organism. EGFP can also be used to mark (or tag) other proteins by creating recombinant DNA constructs that express fusion proteins that fluoresce and can be tracked in living cells or organisms. In this project, we are not using *egfp* as a reporter, but rather as a convenient gene to clone and assay for, as we learn the basic techniques of recombinant DNA manipulation and protein expression.

Reference

1   Yang T, Cheng L, Kain SR. Optimized codon usage and chromophore mutations provide enhanced sensitivity with the green fluorescent protein. *Nucl. Acids Res.* 1996;24(22):4592–4594.

LAB SESSION 1

Getting Oriented: Practicing with Micropipettes

Goal: Starting next week, you will be working on a laboratory project that will build throughout the entire semester. Before embarking on that journey, it is important to familiarize yourself with your lab space and to master the use of the workhorses of the molecular biology lab: the micropipettes. If your instructor has not given safety orientation yet, he or she will do so today.

Station Checklist

It is important to familiarize yourself with the work environment and laboratory equipment before beginning experiments. If the laboratory space which you are working in is shared by other laboratory sections at different times, much of the equipment can be shared. There are certain items, however, such as buffers and sterile disposable items that should not be shared between lab groups. Take a moment to go through your bench, shelves and drawers to identify equipment and reagents. Use the station checklist below and notify your instructor if anything is missing from your station. Items that are indicated as "**per group**" should not be shared between different sets of students on different lab days. Label these items with your initials, lab day and station number. Other items should have the station number only.

STATION CHECKLIST

Station Number _____
Name_____ Name_____
_____ one power supply box
_____ one horizontal DNA minigel apparatus for agarose gels
_____ four micropipettes: P10, P20, P200 and P1000
_____ one box 1000 µl sterile tips **per group**
_____ one box 200 µl sterile tips **per group**
_____ one box 10 µl sterile tips **per group**
_____ one ice bucket (or cooler or styrofoam box for ice)
_____ one box Kimwipes (Kimberly-Clark, Roswell, GA)
_____ one 15 ml and one 50 ml styrofoam test tube rack

Molecular Biology Techniques. DOI: 10.1016/B978-0-12-385544-2.00001-6

_____ one pack sterile snap-cap tubes (17 × 100 mm) for overnight bacterial cultures

_____ one test tube rack for snap-cap tubes

_____ one autoclaved container of 1.7 ml microcentrifuge tubes **per group**

_____ two microcentrifuge tube racks

_____ one pack disposable 10 ml pipettes

_____ one plastic (or electric) pipette pump

_____ one 50 ml graduated cylinder

_____ one 500 ml graduated cylinder

_____ one 2 liter polypropylene beaker

_____ two 1 liter polypropylene bottles, one for distilled water and one for 1X TBE buffer **per group**

_____ one 250 or 500 ml Pyrex orange-capped bottle for melting agarose **per group**

_____ one thermal glove for handling microwaved agarose

_____ labeling tape

_____ permanent ink marker (Sharpie)

_____ one plastic squeeze bottle for 70% ethanol

_____ one plastic squeeze bottle for distilled water

_____ one ring-stand with clamp

_____ one pair blunt-ended forceps

_____ two pairs safety eye glasses or goggles

_____ one cardboard freezer box **per group**

_____ protein polyacrylamide mini gel electrophoresis unit (every other station)

_____ protein mini-transblot assembly (every other station)

_____ vortex mixer

_____ microcentrifuge

_____ bunsen burner

_____ heat block

_____ timer

_____ parafilm

_____ two waste containers: one biohazard and one non-biohazard.

Micropipetting

Micropipettes are the tools used to measure the very small volumes of liquid typically necessary when performing molecular manipulations. We will use four different micropipettes in this course. Each micropipette is accurate to measure a defined range of volume, as shown in the table below (Table 1.1).

Table 1.1	Volume ranges of micropipettes
Micropipette	**Volume range**
P10	0.5–10 μl
P20	2–20 μl
P200	20–200 μl
P1000	200–1000 μl

Setting the micropipettes to the desired volume can be a little tricky at first. It is also common for beginners to confuse the P20 and P200 since they typically use the same pipette tips; therefore, remember to check which micropipette you are using before drawing in solution. Many students accidentally measure 20 µl instead of 2 µl or vice versa because of such mix-ups.

Use Figure 1.1 and the instructions below to help with setting up the micropipettes until you are confident enough to set them on your own.

Follow the instructions below for using the micropipettes.

1. Set the desired volume by holding the pipette in one hand and rotating the dials with the other hand. Do *not* dial past the lower limit 000 or the upper limit (shown on the pipette: 10, 20, 200 or 1000). Familiarize yourself with these settings.
2. Attach a tip to the end of the micropipette. To ensure an adequate seal, press the tip on with a slight twist.
3. Depress the plunger to the first stop. This part of the stroke displaces a volume of air corresponding to that indicated on the dial.
4. Immerse the tip to a depth of 2–5 mm into the liquid to be withdrawn. Immersing the tip to deeper levels will cause liquid to adhere to the outside of the tip, causing errors in measurement.

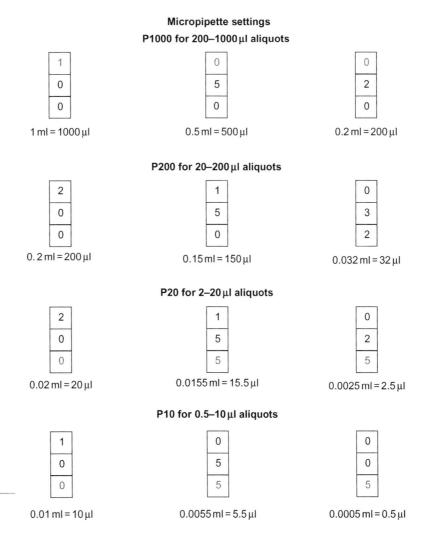

Micropipette settings
P1000 for 200–1000 µl aliquots

1 ml = 1000 µl   0.5 ml = 500 µl   0.2 ml = 200 µl

P200 for 20–200 µl aliquots

0.2 ml = 200 µl   0.15 ml = 150 µl   0.032 ml = 32 µl

P20 for 2–20 µl aliquots

0.02 ml = 20 µl   0.0155 ml = 15.5 µl   0.0025 ml = 2.5 µl

P10 for 0.5–10 µl aliquots

0.01 ml = 10 µl   0.0055 ml = 5.5 µl   0.0005 ml = 0.5 µl

FIG. 1.1
Micropipette settings cheat sheet.

5. Allow the plunger to return slowly to its original position. If the plunger snaps back, aerosols will form contaminating the barrel of the micropipette and your solution.

6. Wait one second before removing the tip from the solution to allow the introduced liquid to enter the pipette tip fully. Removing the tip too quickly from the solution may result in air occupying some of the calibrated volume. Check to make sure that there are no air bubbles and that the amount of liquid corresponds to the desired amount. Develop an eye for 1 μl volumes, as these are the hardest to pipette.

7. Place the tip against the side wall of the receiving vessel near the liquid interface or the bottom of the vessel. Slowly dispel the contents by depressing the plunger until the first stop. Remaining liquid can be dispelled by depressing the plunger to the second stop. Withdraw the tip from the solution and return the plunger to its original position. Check to ensure that no liquid remains in the tip. If there is a bead of liquid, reintroduce liquid from the receiving vessel to capture the bead and slowly expel the contents.

8. Discard the tip by pressing the ejector button.

9. *Always* use a new pipette tip when pipetting enzymes, otherwise the stock solutions may become contaminated. If you accidentally contaminate an enzyme solution, tell an instructor. Always use a new pipette tip for critical volumes, as in a dilution series, because as much as 10% of the volume may stay within the tip after delivery.

10. Working with tiny volumes requires patience and accuracy. The best way to deliver a 1 μl volume is to pick up the receiving tube and make sure that a 1 μl bead is formed on the side of the tube after delivery. In the case of enzymes, schlieren rings should be visible from the glycerol–water interface if the enzyme is dispelled directly into the solution.

Micropipetting Self-Test

Before proceeding further, each student should do a self-test of his or her micropipetting skills. Because 1 ml of water weighs 1 gram, students can test micropipetting skills by pipetting onto a precision balance.

"Passing" the self-test will ensure that you are selecting the correct micropipette for the given volume and that your technique is correct. If your self-test does not fall into the right weight-ranges, see your instructor for one-on-one feedback about your technique, and to test the calibration of your micropipettes.

Each student should perform at least one set of self-tests, selecting and adjusting his or her pipettors independently.

SELF-TEST 1

Volume to measure	Weight (within 5%)
33.5 μl	0.0335
7 μl	0.007
267.5 μl	0.2675

SELF-TEST 2

Volume to measure	Weight (within 5%)
9 μl	0.009
26.5 μl	0.0265
348.5 μl	0.3485

SELF-TEST 3

Volume to measure	Weight (within 5%)
43.5 μl	0.0435
8 μl	0.008
364.5 μl	0.3645

SELF-TEST 4

Volume to measure	Weight (within 5%)
6 μl	0.006
32.4 μl	0.0324
246.5 μl	0.2465

Laboratory Exercise: BSA Serial Dilutions and Nitrocellulose Spot Test

The purpose of this short exercise is to get used to your lab stations and practice using the micropipettes (and to test your technique). Each student will perform serial dilutions of the protein bovine serum albumin (BSA) and then compare their results against their lab partner's results using a visualization technique that uses a protein-binding dye.

Note: While the other laboratory exercises for this course will build on each other, this one will not.

Preparing BSA Dilutions

You will be given a tube with 15 μl of a 1 mg/ml BSA solution. Prepare a dilution series of BSA standards in five tubes (labeled 1–5) according to the scheme outlined in Table 1.2 and Figure 1.2. Make sure to mix each sample before pipetting the next dilution. Attach a new pipette tip to the micropipette each time to make the dilutions. Each lab partner should do a set of dilutions.

Performing a Nitrocellulose Spot Test

Amido black is a stain that quantitatively binds protein. We will use a micropipette to deliver small amounts of the BSA serial dilutions to the nitrocellulose and then stain the nitrocellulose. This will enable you to visualize the relative protein amounts in each sample and provide visual feedback on your pipetting/dilution technique.

Table 1.2 Serial dilution scheme

Tube	Dilution	Protein concentration
1	12.5 µl of stock (1 mg/ml) + 37.5 µl dH$_2$O *This is a 1:4 dilution.*	250 µg/ml
2	25 µl from tube 1 + 25 µl dH$_2$O *This is a 1:2 dilution.*	125 µg /ml
3	25 µl from tube 2 + 25 µl dH$_2$O *This is a 1:2 dilution.*	63 µg /ml
4	25 µl from tube 3 + 25 µl dH$_2$O *This is a 1:2 dilution.*	31 µg /ml
5	25 µl from tube 4 + 25 µl dH$_2$O *This is a 1:2 dilution.*	16 µg /ml

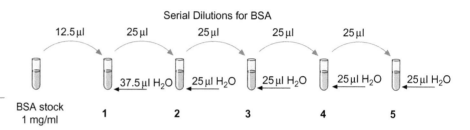

Serial Dilutions for BSA

FIG. 1.2
Serial dilution scheme.

1. Obtain a piece of nitrocellulose (always wear gloves when handling nitrocellulose). Place the nitrocellulose on a piece of 3 MM paper (Whatman, Clifton, NJ) at your station. You will share the piece of nitrocellulose with your partner. If your nitrocellulose membrane is only coated on one side, be sure you use the matte (non-shiny) side. Check with your instructor.

2. Spot 2 µl aliquots of distilled H$_2$O (control) and each of the BSA dilutions. One partner should spot the top row with his/her samples, and the other partner should spot a row below. Spotting of the 2 µl is best done by holding the pipette tip just above the paper. Expel liquid such that a drop forms on the end of the tip. Touch the drop to the paper and the liquid will be drawn into the paper by capillary action.

CAUTION: Make certain you leave enough room between each addition so that the spots do not touch each other.

3. Allow nitrocellulose to air dry.

4. After the spots have dried completely, stain by placing in a tray (a square Petri dish works well for this purpose) and covering with amido black staining solution. Allow to stain for 1–2 minutes with gentle shaking.

5. Pour off the stain (back into original bottle – this can be reused) and cover with methanol-acetic acid destaining solution and shake gently. Change once after 5 minutes and shake gently until the background is white.

6. Place the nitrocellulose on 3 MM paper to dry. Compare the intensities of each spot. Do the intensities of your spots match those of your lab

partner? Does each spot appear to be half as intense as the last? If not, you need to practice your micropipetting technique.

Discussion Questions

1. What are some real-life applications of biotechnology? What are some important recombinant proteins and/or recombinant organisms that are used today?
2. What are your goals in taking this class? What are you hoping to learn, and how do you hope it will expand your career or future research?

LAB SESSION 2

Purification and Digestion of Plasmid (Vector) DNA

Goal: Today you will isolate plasmid DNA. pET-41a is the expression vector you will use for cloning. You will perform the plasmid purification using the QIAprep Spin Miniprep Kit. This protocol starts with an alkaline lysis procedure to break open the cells and separate the plasmid DNA from chromosomal DNA, and is followed by silica adsorption for further purification from soluble cellular proteins and other cellular debris. We will then quantify the DNA.

Introduction to Plasmid Purification

In molecular biology *Escherichia coli* serves as a factory for the synthesis of large amounts of cloned DNA. Today you will isolate plasmid DNA from *E. coli* for *in vitro* manipulation.

Plasmid DNA is cloned in bacteria; that is, identical copies are made and propagated in bacteria. Bacterial cells are a complex mixture of plasmid DNA, chromosomal DNA, proteins, membranes and cell walls. The trick in isolating pure plasmid DNA is to separate it from chromosomal DNA and from the rest of the cellular components.

Alkaline Lysis

The most common method used for separating plasmid DNA from chromosomal DNA is the alkaline lysis method developed by Birnboim and Doly.[1] They exploited the supercoiled nature and relatively small size of plasmid DNA to separate it from chromosomal DNA.

First, cells are broken open under alkaline conditions. Under these conditions, both chromosomal and plasmid DNA are released and denatured (rendered single-stranded). Denatured DNA can reanneal at neutral pH if it is not kept in alkali for too long and if the complementary strands are able to find each other. Since DNA is supercoiled in the bacterial cell, the two halves of the plasmid DNA remain somewhat intertwined during the incubation in alkali and they are in close proximity for reannealing. Because the chromosomal DNA is so large, it remains bound to cellular proteins and lipids, and in the next step it is precipitated out of the solution along with denatured proteins and lipids by addition of potassium acetate.

Molecular Biology Techniques. DOI: 10.1016/B978-0-12-385544-2.00002-8

The precipitated chromosomal DNA and other impurities are usually removed by filtration or centrifugation. RNA is also generally degraded during the alkaline lysis step simply by adding RNase to the buffer. Double-stranded plasmid DNA remaining in solution can then be precipitated by ethanol or can be purified to a higher level either by anion exchange chromatography or by running the sample over a silica membrane. Steps 5–8 of the plasmid DNA purification protocol used today represent the alkaline lysis portion of the purification protocol.

Silica Adsorption

Although many cellular components were removed during alkaline lysis, including the chromosomal DNA, insoluble/denatured proteins and lipids, many cellular proteins and metabolites still remain. Therefore, for high purity, we must further purify the plasmid DNA. The method we will use to accomplish this is through selective adsorption to a silica membrane. Plasmid DNA is selectively absorbed to a silica membrane under optimized high salt conditions. Impurities are washed through, and then pure plasmid DNA is eluted under low salt conditions.

DNA Quantification

DNA quantification is accomplished by reading the absorbance of a known volume of sample at 260 nm. The average extinction coefficient of pure double-stranded DNA is $50 \mu g/ml$. This means that one A260 unit of double-stranded DNA corresponds to $50 \mu g$ of DNA per ml. To assess the purity of a DNA sample, the ratio of the absorbance at 260 nm over the absorbance at 280 nm is calculated. A ratio of approximately 1.8 is ideal. A sample with a higher ratio may have RNA contamination, and a sample with a lower ratio may have protein contamination.

Introduction to Expression Vectors

In general, cloning vectors are plasmids that are used primarily to propagate DNA. They replicate in *E. coli* to high copy numbers and contain a multiple cloning site (also called a polylinker) with restriction sites used for inserting a DNA fragment. A selectable marker, such as an antibiotic resistance gene, is included to select for bacteria containing the plasmid and to ensure its survival. A screenable marker, such as β-galactosidase, is also often included. An expression vector is a specialized type of cloning vector. Expression vectors are designed to allow transcription of the cloned gene and translation into protein. They do have some features in common with the general cloning vectors that are used only for propagating DNA, such as the multiple cloning site and the selectable marker, but they tend to have a lower copy number within cells and they rarely have a screenable marker. They also have some important additional features which allow them to express genes and make protein, including a promoter, ribosome binding site, ATG start codon, a multiple cloning site (polylinker) that allows inserts to be ligated in a predictable reading frame, and often (not always) a fusion tag to aid in purification steps.

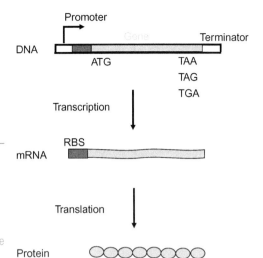

FIG. 2.1

Central Dogma of Molecular Biology. DNA is transcribed into mRNA, which in turn is translated into protein. RNA polymerase binds to the promoter of a gene on DNA and proceeds with transcription, producing a new mRNA. The ribosome and tRNA work together to translate the nascent mRNA into protein.

Principles of Gene Expression

In order to understand how expression vectors function, it is important to recall the Central Dogma of Molecular Biology (Figure 2.1). For a gene to be expressed, it must first be transcribed into messenger RNA (mRNA), and then translated into protein. In the simplest example, RNA polymerase binds to the promoter of a gene and then proceeds with transcription, producing mRNA. Transcription ends at the terminator sequence.

The ribosome then binds the mRNA at the ribosome binding site (RBS) and the ribosome moves along the mRNA. As the ribosome moves along the mRNA, transfer RNA (tRNA) is responsible for decoding the mRNA and specifically depositing an amino acid residue on the nascent polypeptide chain. The translational start codon, which is usually encoded by AUG (ATG on the DNA), encodes the first amino acid (usually methionine), and the translation stop codon (TAA, TAG or TGA) ends translation.

Expression Vectors

The expression vector you will use for your project is pET-41a (Figure 2.2). This expression vector utilizes a kanamycin resistance gene as a selectable marker and the glutathione-S-transferase gene (*gst*) as a fusion tag. The multiple cloning site is downstream (3′) of the *gst* gene and there is no stop codon or termination signal following the *gst* gene. Therefore, when our gene of interest (*egfp*) is cloned into the multiple cloning site it will be expressed as a fusion protein with *gst*, resulting in expression of the fusion protein, GST::EGFP. You will learn more about how creating this fusion protein will aid in the purification of the EGFP protein later in the semester.

The expression of the *gst* gene, and consequently the fusion gene in your future construct, is under the control of the T7 promoter and is inducible using isopropyl-β-D-thiogalactopyranoside (IPTG). In nature, the promoter is induced by lactose, and IPTG mimics lactose with regard to the induction properties, but is not cleaved by the *E. coli* enzyme β-galactosidase. Inducibility is due to the fact that pET-41a uses two

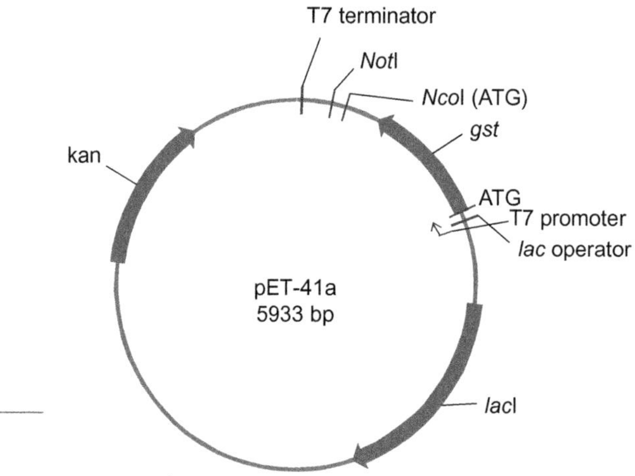

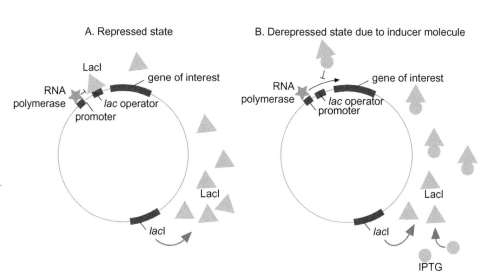

FIG. 2.2

Salient features of pET-41a.

FIG. 2.3

Promoter repression by LacI and derepression by IPTG. (A) The repressed state of the promoter. (B) The derepressed state of the promoter due to the inducer molecule, IPTG.

components of the *lac* operon, the *lac* operator and the *lac*I gene, to regulate transcription. In this vector, the *lac* operator is located adjacent to the T7 promoter. *lac*I encodes a repressor and is constitutively expressed, so the repressor protein LacI is always present. LacI binds to the *lac* operator in the absence of inducer and prohibits RNA polymerase from initiating transcription from the T7 promoter. When the inducer molecule IPTG is added, it interacts with LacI in such a way that LacI will no longer bind to the *lac* operator, and thus transcription by the T7 RNA polymerase proceeds. This process is called derepression of the promoter (Figure 2.3).

Orientation and Reading Frame

When cloning the gene of interest into an expression vector, it is critical for the gene both to be in the correct orientation and the proper reading frame with respect to the start of translation (the ATG start codon encoded by the vector).

Orientation

To ensure that our gene of interest will be inserted in the proper orientation, we will employ the method of directional cloning, also called "forced"

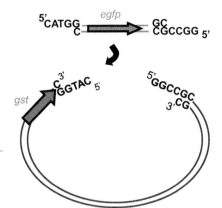

FIG. 2.4

An example of forced cloning using the *Nco*I and *Not*I restriction endonucleases. The insert can only be incorporated in one orientation.

cloning (Figure 2.4). In forced cloning, the polylinker of the vector is digested with two different restriction endonucleases that leave incompatible cohesive ends, and the small "stuffer fragment" between the two restriction sites is excised. The insert (your gene of interest) is then cut out with the same two restriction endonucleases and ligated into the vector. Cloning in this manner, rather than cloning by cutting with only one restriction endonuclease has two advantages:

1. The incompatible cohesive ends will prevent the vector from religating without incorporating the insert (although if the stuffer fragment is not removed, it can be inserted back into the larger portion of the vector instead of the desired insert).
2. The orientation of the insert is forced in a single direction; that is, the 5′ end of the gene can ligate with only one end of the cut vector. Because we know the sequence of both the vector and insert, we know that the fusion protein gene will be transcribed and translated (expressed) correctly.

You will cut the vector with two restriction enzymes that have recognition sequences within the multiple cloning site: *Nco*I and *Not*I. Later, the *egfp* PCR product will be digested using the same two restriction enzymes (*Nco*I at the 5′ and *Not*I at the 3′ end) and ligated into the expression vector.

Remember that the Watson and Crick strands are anti-parallel. If *Nco*I was the **only** restriction enzyme used to cut out both the *egfp* insert and the pET-41a vector, the insert would have the ability to be incorporated into the vector in either orientation. In this case, only 50% of your transformants would contain the *egfp* gene in the proper orientation. Only transcription from DNA in the correct orientation will result in the correct mRNA and the correct amino acid sequence being produced. DNA can only be transcribed in a 5′ to 3′ direction, and the sequence on the bottom strand, 5′ to 3′, is different from the sequence on the top strand.

Reading Frame

The reading frame with respect to the translational start site must be maintained for correct expression. In pET-41a, the junction of foreign DNA with *gst* has to be in the proper reading frame in order to create the desired GST::foreign peptide fusion protein. Most expression vectors are designed in families of three members. Typically, all three expression vectors in the family are identical, except that the reading frames with respect to the multiple cloning sites differ. For example, for a given restriction site, the first vector may put an insert in the +1 reading frame with respect to *gst*, the second in the +2 reading frame and the third in the +3 reading frame. Only one of the three vectors will maintain the correct reading frame for a given insert. The other two will result in the insert being in the wrong reading frame.

pET-41a and several other recently developed expression vectors have an additional feature of the multiple cloning site; an *Nco*I restriction site. The *Nco*I recognition sequence is useful in that it contains an ATG sequence, the start codon for most proteins. The complete *Nco*I recognition sequence is CCATGG. The *Nco*I recognition sequence in pET-41a is located such that the ATG of the sequence is in-frame with the ATG start codon of the *gst* fusion tag. Therefore, if your gene of interest starts with an ATG that is part of an *Nco*I site, then the vector and the 5′ end of the insert (your gene of interest) can both be cut with *Nco*I, and the gene will automatically be in the correct reading frame for translation. Note that the ATG of the gene of interest will not serve as a start codon once ligated into the vector; it will simply encode methionine. The ATG start codon of the *gst* fusion tag found in pET-41a is still used for signaling translation of the fusion protein.

Fortunately, the *egfp* sequence that you will clone into pET-41a does contain the *Nco*I recognition sequence at the ATG start of translation. Therefore, both the vector and the 5′ end of the insert may be cut with *Nco*I and then ligated together, with confidence that the insert will be in the correct reading frame.

Illustrated below is a portion of the multicloning region of the pET-41 family of expression vectors. Note how the *Nco*I site is in the same reading frame for all of the vectors, but addition or deletion of a single base pair downstream of the *Nco*I site changes the amino acid sequence, while also setting up the downstream restriction sites in different reading frames. *Bam*HI is highlighted as a reference site. The restriction site for *Nco*I is **C/CATGG** and the restriction site for *Bam*HI is G/GATCC.

pET-41a: ATG…*gst*…CC**/C-ATG-G**GA-TAT-CGG-G/GA-TCC-GAA-TTC
          Met   Pro Met Gly Tyr Arg Gly Ser Glu Phe

pET-41b: ATG…*gst*…CC**/C-ATG-G**AT-ATC-GGG/-GAT-CCG-AAT-TC
          Met   Pro Met Asp Ile Gly Asp Pro Asn

pET-41c: ATG…*gst*…CC**/C-ATG-G**CG-ATA-TCG-GG/G-ATC-CGA-ATT-C
          Met   Pro Met Ala Ile Ser Gly Ile Arg Ile

Laboratory Exercises

Alkaline Lysis and Silica Adsorption Protocol

The protocol below is modified from the Qiagen QIAprep Miniprep Kit handbook. You will start at step 3.

Note to instructor: If you do not have a Nanodrop available for quantification, students will either need to scale up their preps, or you will need to combine the class preps in order to have a large enough volume to quantify using a standard spectrophotometer, and then aliquot and redistribute DNA for the restriction digestion.

1. Two afternoons before your laboratory, your instructor picked a single colony of *E. coli* strain NovaBlue (or other K12 strain) containing the pET-41a plasmid, from a freshly streaked Luria-Bertani (LB)/kanamycin plate. A starter culture of 2–5 ml LB medium containing kanamycin was inoculated and incubated for ~8 hours at 37°C with vigorous shaking (~300 rpm).

Note: Use a snap-cap tube or flask with a volume of at least four times the volume of the culture to provide adequate aeration.

2. The evening before your laboratory, your instructor diluted the starter cultures 1:500 to 1:1000 into 100 ml selective LB/kanamycin medium. He or she used a flask or vessel with a volume of at least four times the volume of the culture. The culture should reach a cell density of approximately $3–4 \times 10^9$ cells per ml, which typically corresponds to a pellet wet weight of approximately 3 g/liter of medium.

3. Obtain 1.5 ml of the culture in a microcentrifuge tube.

4. Harvest the bacterial cells by centrifugation at 12,000 rpm for 30 seconds. Remove all traces of supernatant by micropipetting. (Note: If you wish to stop the protocol and continue later, freeze the cell pellets at −20°C. Don't do this today, though.)

5. Resuspend the bacterial pellet in 250 µl Buffer P1. The bacteria should be resuspended completely until no cell clumps remain. *Buffer P1 is the resuspension buffer.*

6. Add 250 µl Buffer P2, mix gently but thoroughly by inverting four to six times. Do not vortex, as this will result in shearing of genomic DNA. If necessary, continue inverting the tube until the solution becomes viscous and slightly clear. Do not allow lysis reaction to proceed for more than 5 minutes. *Buffer P2 contains a detergent (sodium dodecyl sulfate; SDS) and sodium hydroxide and is used for cell lysis and denaturation of DNA.*

7. Add 350 µl Buffer N3 to the lysate, mix immediately and thoroughly but gently by inverting four to six times. After addition of Buffer N3, a fluffy white precipitate containing genomic DNA, proteins, cell debris and SDS becomes visible. The buffers must be mixed completely. If the mixture still appears viscous and brownish, more mixing is required to completely neutralize the solution. *Buffer N3 neutralizes the solution, causing plasmid DNA to reanneal, and acts to precipitate the chromosomal DNA and insoluble proteins.*

8. Centrifuge for 10 minutes at 13,000 rpm in a microcentrifuge.

9. Apply the supernatant (liquid above pellet) from step 8 on the QIAprep spin column (containing the silica membrane) by decanting or pipetting. Place column in accompanying tube.

10. Centrifuge for 30 seconds. Discard flow-through.

11. Wash QIAprep spin column by adding 0.75 ml Buffer PE and centrifuging for 30 seconds.

12. Discard flow-through, and centrifuge for an additional 1 minute to remove residual wash buffer.

13. Place the column in a clean 1.5 ml microcentrifuge tube (with the lid cut off). To elute DNA, add 50 μl Buffer EB to the center of each column, let stand for 1 minute, and centrifuge for 1 minute.

14. Your pure DNA is at the bottom of the tube! Label a new tube "pET-41a" and transfer the DNA to the labeled tube.

DNA Quantification

Next, you will quantify the plasmid DNA. Depending on the equipment available in your laboratory, you will either use a Nanodrop or a standard spectrophotometer. Both protocols are given below, although your instructor may make modifications to the spectrophotometer protocol depending on the model of equipment and cuvette size.

OPTION 1: USING THE NANODROP

This method utilizes the Nanodrop. If you have a Nanodrop available, this is the preferred method, both because it is the simplest and because it uses the smallest volume of your precious DNA.

1. Turn on computer and log in if necessary.
2. Open "Nanodrop Software" from the desktop.
3. Click "Nucleic Acid."
4. Wipe sample pedestals (top and bottom) with a Kimwipe, and load 2 μl of water to initialize the machine. *To load a sample, lift arm, pipette the sample on the small silver circle and close the arm back down, making sure the sample contacts both the top and bottom pedestals.*
5. Click "OK" and the Nanodrop will click a couple of times.
6. Lift the arm and wipe off water sample from top and bottom with a Kimwipe.
7. Load 2 μl of your blank (Buffer EB or TE buffer), close arm, and click "Blank," which is in the upper left-hand corner of the window on the computer screen. The Nanodrop will click.
8. Wipe off blank from pedestal top and bottom with a Kimwipe.

FOR YOUR SAMPLE

9. Load DNA sample (2 μl), and click "Measure," in the upper left-hand corner of the window. The Nanodrop will click as it measures the sample.
10. The A260, A280, A260/A280 and the concentration will appear in the bottom right-hand corner of the window. Record all of these numbers in your lab notebook.
11. Wipe off the sample from top and bottom, and repeat process for all other samples – blanks do not need to be repeated. **If there is a several-minute gap between sample readings, a re-blanking is advised.**
12. Label your tube of purified DNA with your lab day and station number, the name of the plasmid you purified and the concentration (with the units ng/μl). Keep this sample on ice (and save remainder in freezer box at the end of the day).

OPTION 2: USING THE STANDARD SPECTROPHOTOMETER

This method for DNA quantification uses the GeneQuant apparatus from Amersham Biosciences. It also uses a special cuvette that accepts a quantity as small as 70 μl. This cuvette allows for minimal sample waste. If this apparatus is not available, a standard uv/vis spectrophotometer can be used to assess the absorptions at 260 and 280 nm.

Note to instructor: If you are using this method, you may either combine student preps, or have had each student scale-up to a midiprep or multiple minipreps. Using 1.5 ml of starting culture in a single miniprep, students typically obtain a concentration of 50 ng/μl. Therefore, diluting the sample is not advised.

1. Turn on the spectrophotometer 15 minutes before use.
2. Blank the spectrophotometer using 70 μl of Buffer EB in the quartz cuvette. Use non-powdered gloves. Be especially careful not to get fingerprints on the clear side of the cuvette. If you think you left fingerprints, rinse and wipe well with a Kimwipe.
3. After the instrument has been blanked, carefully empty and rinse the cuvette. Please be careful with the quartz cuvette. These cuvettes are shockingly expensive and are NOT disposable.
4. Read the absorbance of 70 μl of your DNA at 260 nm and 280 nm. Record the readings in your notebook.
5. Empty and rinse cuvette.
6. Calculate the concentration and purity of your original sample.
7. To determine the concentration of your DNA, use the equation: (A260)×(50 ng/μl)×(dilution factor) = DNA (ng/μl) Remember, in this case, your dilution factor was 1 (undiluted).
8. To determine the purity of your DNA, calculate the ratio: A260/A280.
9. Record the DNA concentration and ratio in your notebook. If the A260/A280 ratio was significantly different from 1.8, see your instructor.
10. Label your original tube of purified DNA with your station number, pET-41a, the concentration (with the units ng/μl) and the purity. Keep this sample on ice (and save remainder in freezer box at the end of the day).

Restriction Digestion of Expression Vector DNA pET-41a, a GST Fusion Protein Vector

Goal: You will prepare the expression vector plasmid, pET-41a (Novagen), to be able to accept the gene (*egfp*) you are going to clone in the following weeks. To accomplish this, you will digest pET-41a simultaneously with two restriction endonucleases, *Nco*I and *Not*I. This will allow you to clone the *egfp* gene into the vector in a single orientation, ensuring correct translation of a GST::EGFP fusion protein.

RESTRICTION ENZYME DIGESTIONS

Restriction enzyme activity is defined as the amount of enzyme (measured in units, U) that will cleave 1 μg of DNA (usually λDNA) to completion in 1 hour at the optimum temperature for the enzyme, usually 37°C. Buffers are usually supplied with restriction enzymes at a 10× concentration. As a general rule, to set up a restriction enzyme digestion:

• determine the amount of DNA to be cleaved;
• use a five-fold excess of enzyme;
• ensure that the volume of enzyme does not exceed 10% of the final volume;

19

- add 10× buffer to a final concentration of 1×;
- enzymes should be added to the reaction last.

Some enzymes will cleave at a second site under sub-optimal conditions, producing what is referred to as "star activity." Each group needs to digest 1000 ng (1 μg) of pET-41a vector. To determine what volume of your DNA to add to the digest, use the following equation:

Volume to add (μl) = Amount you want (1000 ng)/Concentration (ng/μl)

1. Label a fresh tube for your restriction digestion with your lab day and station number (i.e. "W9") and "pET digest."
2. Digest 1 μg (1000 ng) of pET-41a DNA by adding the reagents listed below to the labeled tube, being sure to add the reagents in order. Centrifuge tubes that contain small volumes of liquid for 5 seconds before removing aliquots – enzyme, DNA, buffer, etc. Make sure that buffers, which are stored at −20°C, are completely thawed and vortexed before using.

Note: This and other protocols use enzyme and buffer from New England Biolabs. Other brands of restriction endonucleases may be used but be sure to use the buffer suggested by that particular manufacturer at the concentration suggested by the manufacturer. Much of this information is available on the web.

   ___ μl pET-41a DNA (as calculated based on your concentration)
   ___ μl dH$_2$O (as calculated to achieve 50 μl total volume in reaction)
   5 μl 10× restriction buffer (NEB buffer 3)
   0.5 μl BSA
   2 μl *Nco*I (always add enzyme last)
   2 μl *Not*I (always add enzyme last)

3. Mix and then spin for 2–5 seconds to bring the contents to the bottom of the tube.
4. Place tube in a microfuge rack in a 37°C incubator for at least 80 minutes. It is critical for the digestion to go to completion. At the completion of the digestion, store your sample in your freezer box (in the freezer!) along with the tube of the remainder of your uncut plasmid. **You will be using both your digested and your uncut vector in the future, so it is critical to save them at −20°C.**

Discussion Questions

1. If you do an absorbance reading after plasmid purification and get a A260/A280 of less than 1.8, how could you further purify the sample to get rid of the protein contamination? Is it always necessary to have completely pure DNA? What are some cases where it would or would not be?
2. Why do we increase the pH to denature the plasmid and chromosomal DNA during alkaline lysis rather than using high temperatures, which would also denature DNA?

Reference

1   Birnboim HC, Doly J. A rapid alkaline extraction procedure for screening recombinant plasmid DNA. *Nucleic Acids Res.* 1979;7:1513.

LAB SESSION 3

PCR Amplification of *egfp* and Completion of Vector Preparation

Introduction

The gene we will be cloning and expressing is *egfp* (the gene encoding the enhanced green fluorescent protein). The green fluorescent protein (GFP) is a naturally occurring protein found in a species of fluorescent jellyfish called *Aequoria victoria*. The difference between the fluorescence of the green fluorescent protein (GFP) and the enhanced green fluorescent protein (EGFP) is that EGFP emits 35× the fluorescence of GFP when excited with ultraviolet or blue light: it is much brighter. The increased fluorescence was achieved by making mutations in the nucleic acid sequence that resulted in a small change in the amino acid composition within the chromophore region of the protein: Ser65→Thr and Phe64→Leu.[1] To clone the *egfp* gene, we will use a PCR-based strategy.

Cloning by PCR is one of many techniques that utilize a PCR-based strategy. In this manual alone, we will employ PCR in four different experiments: cloning by PCR; screening for positive clones using anchored PCR; quantitative (real-time) reverse-transcription PCR and semi-quantitative reverse-transcription PCR. There are many other applications that utilize this technique that are beyond the scope of this manual, but may be introduced in your lecture, such as site-directed mutagenesis using a PCR-based strategy. Before proceeding, you should be familiar with both PCR and PCR-based cloning.

What is the Polymerase Chain Reaction (PCR)?

In order to understand the PCR cloning started in this laboratory session, a general understanding of the polymerase chain reaction is required. The polymerase chain reaction (PCR) was developed by Kary Mullis in 1983.[2] It has simplified many procedures in molecular biology and made possible countless new techniques.

PCR uses a logarithmic process to amplify DNA sequences. A thermostable DNA polymerase is used in repeated cycles of primer annealing, DNA synthesis and dissociation of duplex DNA to serve as new templates. The

Molecular Biology Techniques. DOI: 10.1016/B978-0-12-385544-2.00003-X

FIG. 3.1

Orientation of PCR primers in relation to target DNA. The forward primer anneals to the 3′ end of the bottom strand. When the forward primer is extended, a copy of the top strand is created. The reverse primer anneals to the 3′ end of the top strand. When the reverse primer is extended, a copy of the bottom strand is created. If the top strand corresponds to the sense strand, the forward primer creates a copy of the sense strand, even though it binds to the 3′ end of the antisense strand. By convention, the sequence of a gene refers to the mRNA-like strand. The template used by RNA polymerase during transcription is the antisense strand of a gene. This convention makes it easier to conceptualize sequence domains and correlate them with protein motifs.

theoretical amplification of template DNA, assuming reagents are not limiting and the enzyme maintains full activity, is 2^n where n is the number of cycles. After 30 cycles of PCR from a single template, 1×10^9 new DNA molecules could be synthesized. A typical PCR thermocycling program consists of the following steps:

1. denature DNA (94°C) ~1 minute;
2. anneal primers to template (based on the melting temperature of the primers ~60°C for a typical 20-mer) ~1 minute;
3. synthesize DNA (72°C) ~1 minute per kb to be amplified;
4. repeat steps 1–3 thirty times.

DNA primers are short, single-stranded sequences complementary to the ends of the sequence to be amplified and are oriented in opposite directions. In other words, the two primers must flank a DNA region, with one primer annealing to the sense strand and one to the antisense strand, with the primers facing inwards toward each other (Figure 3.1). Both primers are necessary for exponential amplification to occur.

Primers are chemically synthesized on an instrument called an oligonucleotide synthesizer. The annealing temperature of the primers can be estimated from the formula for the melting temperature (T_m) of DNA molecules shorter than 50 bp: $T_m = (4)(\text{number of GC pairs}) + (2)(\text{number of AT pairs})$.

The thermostable DNA polymerase most commonly used is *taq* DNA polymerase, isolated from the thermophilic bacterium *Thermus aquaticus*, which was discovered growing in hot springs at 75°C at Yellowstone National Park. In our experiment we will be using an alternative thermostable DNA polymerase called Vent from *Thermococcus litoralis*. Vent is often used when a low error rate is required because it has 3′→5′ exonuclease (proofreading) activity.

The ingredients necessary for the polymerase chain reaction to take place are:

- template DNA
- forward primer
- reverse primer
- nucleotides (dATP, dCTP, dTTP, dGTP)
- thermostable DNA polymerase
- buffer
- magnesium (necessary for enzyme activity).

For an excellent review of PCR, visit the website http://www.dnalc.org/shockwave/pcranwhole.html.[3]

Why Clone by PCR?

Cloning using a PCR-based strategy is quickly replacing traditional cloning for many applications. Traditional cloning is where the DNA to be cloned is cut out of a larger DNA molecule using restriction enzymes, and then the desired DNA fragment is purified from an agarose gel. There are numerous advantages of cloning by PCR over traditional cloning, including:

1. There is no reliance on having nature provide restriction sites flanking the region you want to clone. In PCR cloning, restriction sites can be engineered into PCR primers to create flanking restriction sites, or other overhangs suitable for cloning can be created.
2. In traditional cloning, if the DNA molecule you are isolating your DNA from is very large (a genome, for instance), you cannot simply separate your DNA fragment using agarose gel electrophoresis in order to purify it since there would be too many other fragments of the same or similar size.
3. PCR is fast!

The main disadvantage of PCR cloning versus traditional cloning is that PCR has a higher error rate than replication *in vivo*, and therefore mistakes are more likely to be incorporated into your cloned gene. To overcome this, we do two things:

1. We use a DNA polymerase that has a proofreading function (such as Vent DNA polymerase).
2. After we find positive clones by screening methods, we always sequence positive clones to check for sequencing errors that could affect the amino acid sequence of the recombinant protein.

There are multiple methods of cloning PCR products. Two of the most commonly used methods are TA cloning and cloning by incorporation of restriction sites into primers.

TA Cloning

One commonly used method, called TA cloning, takes advantage of an unpaired A-overhang which occurs at the 3′ ends of the PCR product in reactions where a non-proofreading DNA polymerase is used. These PCR reactions can be easily ligated into a vector that has been cut open with an enzyme that leaves blunt ends, and then modified to achieve a single T overhang. The downside of this method is that DNA polymerases that do provide a proofreading function (and therefore lower error rate) do not create an A-overhang.

PCR Cloning by Incorporation of Restriction Sites

The PCR cloning method we will employ involves engineering restriction sites into the PCR primers, so that those restriction sites will be incorporated into the PCR product to be cloned. Figure 3.2 depicts a PCR reaction where restriction sites are engineered into the primers. The restriction sites are not encoded by the template DNA, so notice that the sites do not

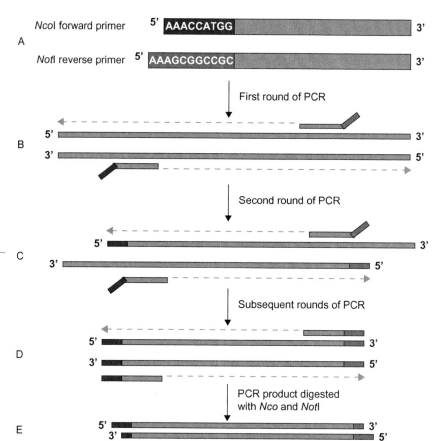

FIG. 3.2

Incorporation of restriction sites into a PCR product by engineering them into the forward and reverse primers. (A) shows in detail the primers with the engineered restriction sites and additional nucleotides necessary for digestion of the PCR product. (B) and (C) show the first two rounds of PCR. Notice the engineered portions of the primers are not complementary to the template and do not anneal. In (D), the third and subsequent rounds of PCR, the primers do fully anneal to the newly synthesized template. (E) depicts the digested DNA with "sticky" 5' ends.

anneal to the template. For this reason, when considering the annealing temperature used in the PCR reaction, we do not add in the T_m coming from the non-complementary portion of the primer. Note that we must also add a few additional nucleotides 5' of the restriction sites on each primer. This is because our next step will be to digest the PCR product for insertion into the vector, and most restriction endonucleases cannot cut restriction sites that are at the very end of the DNA molecule.

Once we have assessed that our PCR reaction successfully yielded a single PCR product of the expected size, we will digest the PCR product with the restriction enzymes NcoI and NotI in order to have compatible cohesive ends with the digested pET-41a expression vector, and proceed to ligation (in later lab exercises).

Cloning Synthetic Genes

An exciting new cloning strategy that is on the horizon, but is still cost-prohibitive for regular use, is the cloning of synthetic genes. In this method, the entire DNA sequence to be inserted into the vector is created *de novo*. This has the advantage that the gene sequence can readily be altered to accommodate differences in codon usage between the organism of origin of the gene and the recombinant host. It can also be used to make other site-directed mutations. In the next several years, as technology costs come down, this method may become more routine.

Laboratory Exercises

PCR Amplification of *egfp* from the pEGFP-N1 Plasmid

Goal: pEGFP-N1 contains the *egfp* gene (enhanced green fluorescent protein gene). You will use PCR to replicate many copies of the gene. The PCR primers you are using are specially engineered to have restriction sites incorporated into them, so that when your PCR reaction is complete, you will be able to digest the PCR product so that it will have sticky ends for cloning.

The forward primer (egfpNco: AAACCATGGTGAGCAAGGGCGA) you will use will incorporate an *Nco*I restriction site on the 5′ end of your PCR product, and the reverse primer (egfpNot: AAAGCGGCCGCTTTACTTGTACA) will incorporate a *Not*I restriction site on the 3′ end of your PCR product.

PCR Protocol

In a PCR tube, mix the following in order:

> __ μl dH$_2$0 (to bring volume to 50 μl)
> 5 μl 10× thermopol reaction buffer
> 100 ng purified pEGFP-N1 plasmid (calculate the volume needed based on the concentration)
> 5 μl dNTP mix (stock of 2 mM each dNTP)
> 1 μl egfpNco primer (100 pmol/μl stock)
> 1 μl egfpNot primer (100 pmol/μl stock)
> 0.5 μl Vent polymerase.

Mix by flicking the tube, then tap reaction down to the bottom. Place reaction in thermocycler and run the following protocol using a heated lid:

1.	Denature	95°C	2 minutes
2.	Denature	95°C	30 seconds
3.	Anneal	60°C	30 seconds
4.	Extend	72°C	1 minute
5.	Repeat steps 2–4 thirty times		
6.	Extend	72°C	5 minutes
7.	Hold	4°C	indefinitely

Your instructor will store your PCR reaction in the −20°C freezer for you until next week. We will now back up and finish preparing our vector for future ligation.

Clean-up of Digested pET-41a Vector

Goal: To remove restriction enzymes, salts and other impurities from digested pET-41a vector.

The vector you linearized in the previous lab will be used in a later lab for ligation with the insert (*egfp*) DNA. Ligations are very sensitive to salt concentrations, so it is important to remove the salts present in the restriction digestion buffer.

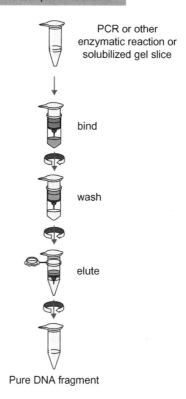

The QIAquick Procedure

PCR or other enzymatic reaction or solubilized gel slice

bind

wash

elute

Pure DNA fragment

FIG. 3.3
QIAquick PCR Purification Kit Flowchart.[4]
Copyright 2008 Qiagen Corporation. Used with permission.

We will use the Qiagen QIAquick PCR Purification Kit Protocol (Figure 3.3) to remove the salts and enzymes. The spin column works on the principle of silica adsorption, just like the column used during the small-scale plasmid prep (mini-prep) previously. Alternatively, one could achieve the same goal by performing an ethanol or isopropanol precipitation, but the QIAquick protocol is preferred because it is more rapid and more efficient (less sample is lost) than an alcohol precipitation.

Protocol: The protocol below is modified from the manufacturer's handbook (QIAquick Spin Handbook 03/2008).

Note: All centrifugation steps are performed at ~13,000 rpm in a conventional microcentrifuge. Balance your tube(s) with one from another lab group.

1. Add 250 μl Buffer PB to your *NcoI/NotI* digested vector from last week and mix.
2. Place a QIAquick spin column in the 2 ml collection tube.
3. To bind DNA, apply the entire sample to the QIAquick column and centrifuge for 30 seconds.
4. Discard the flow-through. Place the QIAquick column back in same collection tube.
5. To wash, add 0.75 ml Buffer PE to the QIAquick column, and centrifuge for 30 seconds.
6. Discard flow-through and place the QIAquick column back in same collection tube. Centrifuge an additional 1 minute.

7. Place the QIAquick column in a clean, sterile 1.5 ml microcentrifuge tube. Cut the lid off the 1.5 ml tube with scissors so that it will fit in the microcentrifuge.

8. To elute the DNA, add 50 µl Buffer EB to the center of the QIAquick membrane and centrifuge for 1 minute.

9. Transfer your purified DNA to a fresh microcentrifuge tube labeled with "pET-41a digest" and your initials.

10. Nanodrop to determine concentration and record (as previously described).

Agarose Gel Electrophoresis

Agarose gel electrophoresis works on the premise that all DNA is negatively charged. When DNA is added to the negatively charged pole (cathode) of an electrical field, it will move toward the positively charged pole (anode). Because of the pore size of the agarose matrix, large DNA molecules will be retarded in the matrix to a greater degree than small molecules, so the small molecules will migrate more rapidly through the gel. Distinct bands ordered according to size can be visualized if the electric field is stopped before bands run off the end.

You will run the digested vector DNA on an agarose gel to visualize whether the DNA cut to completion. Linear DNA migrates as a single band of a predictable size on an agarose gel. Circular plasmid DNA can appear as multiple bands and does not migrate at the same rate as linear DNA because of its secondary structure. In this exercise, you will compare the appearance of the linearized DNA from your reaction to an undigested sample. If your sample is not completely digested, there will be two or more bands instead of one. Make sure that you do not see the circular DNA in the digested sample to ensure that the digested DNA cut to completion. If you do have two or more bands, consult your instructor.

Ethidium bromide (EtBr) was traditionally used to visualize DNA on a gel. EtBr intercalates into double-stranded DNA and then emits orange/pink fluorescence when excited by UV light. However, EtBr is a carcinogen because if it enters cells it can intercalate into DNA causing thymidine dimers, which could lead to mutation. GelRed detects DNA in an analogous way, but has been shown to be unable to cross the cell membrane and thus doesn't pose as great a health risk.

You will each receive a stock of 10,000× GelRed Nucleic Acid Stain (Phenix Research Products). Note: This tube should be saved and stored protected from light. You will use this stock of GelRed throughout the semester.

1. To prepare a 1% agarose gel, add 1 g of agarose to 100 ml of 1× TBE buffer. Microwave for 30 seconds with the cap on loosely. Swirl and repeat until the agarose is completely in solution. Be sure to use rubber "hot-hands" to handle the hot bottle, and never swirl the bottle close to your face in case the liquid boils over.

CAUTION: Never leave the agarose solution unattended when using the microwave. The solution must be swirled occasionally during the heating process to prevent superheating of local areas. Always use "hot-hands" or autoclave gloves when heating the agarose.

27

2. Pour 30 ml of your melted agarose into a disposable 50 ml conical test tube.

3. Add 3 μl GelRed to the 30 ml agarose and mix thoroughly without creating bubbles. Then pour into the casting tray as demonstrated by your instructor. *Allow the agarose to cool slightly before pouring the gel, because steaming-hot agarose can warp the gel apparatus.*

4. Place the comb in the gel. Allow the remaining agarose to solidify in the bottle with the cap tightly closed. This agarose will be re-melted in the microwave for pouring gels during the next several exercises.

5. While the gel is solidifying, begin preparing your samples.

For sample preparation below: do not add loading dye to the entire original sample; if you do, it can't be used for cloning.

6. To prepare your digested DNA sample for loading on the gel, mix together in a separate tube:
   __ μl digested pET-41a (to equal 50 ng)
   __ μl sterile distilled water (to bring to 10 μl if necessary)
   1 μl 10× loading dye.
   Mix, then centrifuge for 2 seconds.

7. To prepare uncut DNA for loading on the gel, mix together in a separate tube:
   __ μl uncut pET-41a (to equal 50 ng) (save the rest in freezer box for a later laboratory)
   __ μl sterile distilled water (to bring to 10 μl if necessary)
   1 μl 10× loading dye
   Mix, then centrifuge for 2 seconds.

8. Once the gel has completely solidified, pour about 160 ml 1× TBE over the gel, remove dams and comb.

9. Load your gel as follows:
   Lane 1: empty;
   Lane 2: 10 μl NEB 1 kb DNA ladder, premixed with loading dye;
   Lane 3: your digested DNA sample described above;
   Lane 4: empty;
   Lane 5: uncut DNA described above.

The NEB 1 kb DNA ladder is a molecular weight marker for estimating DNA size. The visible bands are approximately 10, 8, 6, 5, 4, 3, 2, 1.5, 1 and 0.5 kilobases in size (Figure 3.4).

10. Once you have loaded the samples, close the apparatus and connect the leads to the voltage pack, as demonstrated by your instructor. Run your DNA gel at approximately 85 volts, or at an appropriate voltage for your apparatus (10 V/cm). Once the bromophenol blue dye-front runs to the bottom half of your gel, stop the current and place the gel on a UV transilluminator. Wear UV protective goggles whenever direct exposure to UV is a possibility. Photograph the gel and examine the photograph. Your instructor will help you determine whether your plasmid cut to completion. There should be a single DNA band in your cut DNA sample lane, and it should run at a slightly different size than uncut DNA.

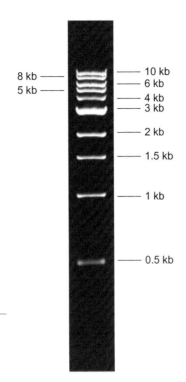

8 kb ———
5 kb ———

——— 10 kb
——— 6 kb
——— 4 kb
——— 3 kb

——— 2 kb

——— 1.5 kb

——— 1 kb

——— 0.5 kb

FIG. 3.4

NEB 1 kb DNA ladder (catalog number N3232).
Note that the 3 kb band is brighter than the
adjacent bands.

Discussion Questions

1. How efficient would PCR be if we set the annealing temperature higher or lower than the calculated melting temperature? How would a higher or lower temperature affect the annealing capability of the primers and the final quantity or quality of the products? How would a gel of such PCR products look compared to PCR products obtained at the optimum annealing temperature?

2. People are most familiar with using *taq* DNA polymerase but in this lab we are using a different thermostable DNA polymerase. What is it and why are we using it instead?

3. Why can circular plasmid DNA appear as multiple bands on an agarose gel? Why doesn't it run at the same apparent size as linear DNA of the same length on the gel?

References

1   Yang T, Cheng L, Kain SR. Optimized codon usage and chromophore mutations provide enhanced sensitivity with the green fluorescent protein. *Nucl. Acids Res.* 1996;24(22):4592–4594.

2   Mullis KB. The unusual origin of the polymerase chain reaction. *Sci. Am.* 1990;262(4): 56–65.

3   Dolan DNA Polymerase Chain Reaction. Learning Center: Biology Animation Library Cold Spring Harbor Laboratory.

4   QIAGEN® QIAquick Spin Handbook. March 2008. <http://www1.qiagen.com/ literature/handbooks/PDF/DNACleanupAndConcentration/QQ_Spin/1021422_ HBQQSpin_072002WW.pdf/>

LAB SESSION 4

Preparation of Insert DNA (*egfp*) PCR Product

Goal: Today you will confirm the success (hopefully!) of your PCR reaction and digest your *egfp* PCR product to create *NcoI/NotI* sticky ends to be used for ligation next week. Review Figure 3.2 prior to this laboratory.

Check PCR Reactions on an Agarose Gel

Goal: To determine whether your PCR reaction was successful in amplifying the 724 bp *egfp* gene.

PROTOCOL

(Review Lab Session 3 section on Agarose Gel Electrophoresis.)

1. Pour a 1% agarose gel as described previously (steps 2–4 should be performed while waiting for the gel to solidify and while the gel is running).
2. Remove 5 μl of your PCR product to a new tube (saving the remainder on ice).
3. Add 1 μl DNA loading buffer to the 5 μl aliquot.
4. Load your gel as follows:
   - 10 μl 1 kb ladder (saved from last time);
   - 6 μl of your sample mixed with loading dye.
5. Run and visualize gel as previously described.

Spin Column Cleanup of PCR Product

You may begin this while you are waiting for your gel to solidify (after removing the 5 μl to run on the gel).

1. Perform Qiagen QIAquick PCR Purification Kit Protocol to remove the salts, enzyme and unincorporated nucleotides from your PCR product, as previously described in Lab Session 3 (cleanup of digested pET-41a vector).

Quantification of *egfp* PCR Product

1. Use the Nanodrop, as previously described, to determine the concentration of your undigested *egfp* PCR product. If no Nanodrop is available, proceed directly to the restriction digestion, using 20 μl of *egfp* PCR product.

Molecular Biology Techniques. DOI: 10.1016/B978-0-12-385544-2.00004-1

Restriction Digestion of *egfp* PCR Product

You PCR amplified the *egfp* gene using primers that have the restriction sites *Nco*I and *Not*I engineered onto the ends (with a few additional nucleotides). Right now, the *egfp* gene still has blunt ends from the PCR reaction. We need to digest the DNA in order to create the sticky ends for cloning. To accomplish this, you will digest your *egfp* PCR product simultaneously with two restriction endonucleases, *Nco*I and *Not*I. This will allow you to clone the *egfp* gene into the vector in a single orientation, ensuring correct translation of a GST::EGFP fusion protein.

RESTRICTION DIGEST PROTOCOL

(This should be done WHILE running your agarose gel.)

1. Digest the *egfp* PCR product by adding the reagents listed below, being sure to add reagents in order using a fresh pipette tip for each ingredient. Centrifuge tubes that contain small volumes of liquid for 5 seconds before removing aliquots – enzyme, DNA, buffer, etc. Make sure buffers that are stored at −20°C are completely thawed before using.

Remember: Volume to add (μl) = Amount you want (500 ng)/Concentration of sample (ng/μl).

In a tube labeled with your lab day, station number and "egfp digest," add the following in order:
__ μl *egfp* PCR product to equal 500 ng (save the rest in freezer box stored at −20°C)
__ μl dH$_2$O to total volume of 50 μl
5 μl 10× restriction buffer (NEB buffer 3)
0.5 μl BSA
1 μl *Nco*I (always add enzyme last)
1 μl *Not*I (always add enzyme last).

2. Mix and spin (with another tube used as a balance) for 2–5 seconds to bring contents to the bottom of the tube.
3. Place tube in a microfuge rack in a 37°C incubator for at least 1 hour.

Removing Enzymes and Cleaning Digested DNA Using a Spin Column

Your *egfp* gene now has sticky ends for cloning. Ligations are very sensitive to salt concentrations, so it is important to remove the salts used in the restriction digestion buffer. Perform the Qiagen QIAquick PCR Purification Kit Protocol once again on the digested PCR product, and then check the concentration by Nanodrop.

If no Nanodrop is available, run 10 μl of your purified digested PCR product alongside the 1 kb ladder on an agarose gel. Your instructor can help you estimate your DNA concentration comparing against the mass amount in the ladder. Label this tube very carefully with your lab day, station number and "clean egfp dig" and store in your freezer box with your clean, digested vector for next week.

Calculations for next week (you must complete these in lab today and have them checked!).

- Calculate # of μl of your digested pET-41a to equal 50 ng.
- Calculate # of μl of your clean, digested *egfp* insert to equal 18 ng.

Use the equation:

Volume needed (μl) = Amount desired (ng)/Concentration of sample (ng/μl)

Discussion Questions

1. What are the potential reasons why more than one band may appear when you run your PCR product on a DNA gel?
2. Would the cloning of the *egfp* gene into the pET-41a vector work if we skipped the *NcoI*/*NotI* digestion of the PCR-amplified *egfp*? How would the results of next week's ligation be affected?
3. Normally, a few additional nucleotides need to be engineered 5′ of the restriction sites on both the forward and reverse primers when cloning by PCR in the method we are using. Why is this necessary? How can you determine how many additional nucleotides should be added to the ends of the PCR primers? Does it matter which nucleotides are added?

LAB SESSION 5

DNA Ligation and Transformation of *Escherichia coli*

Goal: The goal of this week's laboratory is to ligate the *egfp* gene (our insert) into the linearized pET-41a expression vector. We will then transform the ligation mixture into *E. coli* host cells.

Introduction

Ligation

DNA ligase catalyzes covalent bond formation between the 3′OH and 5′PO$_4$ on DNA (phosphodiester bond). It requires two ends of double-stranded DNA and a buffer that includes ATP. Ligation kinetics are complex and it is worthwhile for you to understand some of the parameters that affect both the frequency and products of ligation. Bacteriophage T4 DNA ligase is the preferred enzyme for DNA ligation because it can efficiently ligate blunt-ended DNA, as well as DNA with compatible cohesive ends (short, complementary 5′ or 3′ extensions of single-stranded DNA created by restriction enzyme digestion).

The optimal ligation temperature for a given reaction has been shown empirically to vary. DNA with compatible cohesive ends is the most amenable to ligation, although blunt ends of DNA can also be ligated together.

Different preparations of ligase and ligase buffer call for differing ligation times and temperatures. Ligation reaction temperatures are a balance between the activity of the enzyme and the annealing temperature of the compatible ends. For this reason, molecular biologists usually perform ligation reactions with compatible cohesive ends at room temperature, while we perform blunt-ended ligation reactions at 4°C. Always be sure to check with manufacturers' instructions. When designing ligation reactions, there are several practical rules of thumb that are useful to follow:

- Ligation reactions are generally set up in small volumes of 10–20 μl so that compatible (or blunt) ends will not be too dilute in solution.
- For average-sized vectors (2–6 kilobases), use approximately 50 ng of vector DNA per reaction. Dugaiczyk (1975)[1] empirically derived a formula for determining how much vector is ideal in a ligation reaction.

Molecular Biology Techniques. DOI: 10.1016/B978-0-12-385544-2.00005-3

- The ideal amount of insert to use can only be determined empirically, but common vector-to-insert molar ratios used are 1:1, 1:3 and 1:10.
- Ligase buffer includes the necessary salts and ATP, and is usually supplied as a $10\times$ solution by the ligase manufacturer. Be sure to include it in your ligation mix or the ligase will not be active.
- $1\,\mu l$ ligase per reaction from most commercially available sources is sufficient; however, a more concentrated ligase preparation may be useful for blunt-end ligations.

For today's experiment, we will attempt to ligate the expression vector with the *egfp* insert using a 1:3 vector to insert ratio. In addition to our experimental ligation and transformation, we will also perform controls to confirm the viability and competence of our cells, as well as to confirm that our vector did indeed cut to completion in Lab Session 2. While we previously looked at digested pET-41a on an agarose gel to determine whether the plasmid was linearized, the ligation/transformation control is more sensitive. There are two pieces of information that are critical to understanding these controls.

1. DNA strands do not spontaneously ligate, even if the strands have complementary cohesive ends. An enzyme that catalyzes phosphodiester bond formation (such as DNA ligase) **must** be present for ligation to occur.
2. Only circular DNA is transformable into competent *E. coli* cells. Linear DNA can get into the cells, but it cannot replicate once taken up.

Transformation

Bacterial cells can be made competent for DNA uptake by treatment with divalent cations, and can be stored in this state indefinitely at $-80°C$. Once chilled, the cells must not warm to room temperature or they will lose competence. For optimal transformation efficiencies, the cells should be kept on ice.

Because the expression vector we are using, pET-41a, has a T7 promoter that drives transcription of the gene of interest, the *E. coli* host strain must be a λ lysogen so that it will have a T7 RNA polymerase to bind at the promoter for expression. A suitable strain is BL21(DE3).

Figure 5.1 shows outcomes of the two controls and the experimental ligation and transformation. The top row shows the control for viability and competence. The second row shows the control for complete restriction digestion of the pET-41a vector, and the bottom row shows the experimental ligation (vector plus insert). The first column shows the type of DNA to be added to the reaction: circular or linearized. The second column indicates whether DNA ligase is added to the reaction. The third column shows whether that DNA should transform or not (circular DNA can transform, linear cannot). If the plasmid transforms, then transformed cells should grow on a medium containing kanamycin since the vector contains the kanamycin resistance gene. The last column shows some explanations for unexpected results (where colonies grew when they shouldn't have, or vice versa). Keep in mind that restriction digestions and ligation reactions rarely go to completion. It is normal to see a few background colonies on

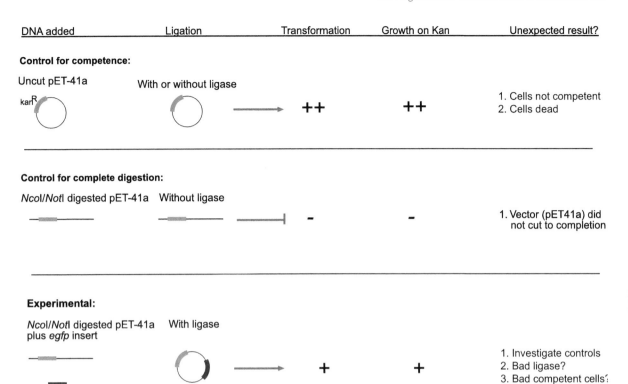

DNA added	Ligation	Transformation	Growth on Kan	Unexpected result?
Control for competence:				
Uncut pET-41a	With or without ligase	++	++	1. Cells not competent 2. Cells dead
Control for complete digestion:				
NcoI/NotI digested pET-41a	Without ligase	-	-	1. Vector (pET41a) did not cut to completion
Experimental:				
NcoI/NotI digested pET-41a plus *egfp* insert	With ligase	+	+	1. Investigate controls 2. Bad ligase? 3. Bad competent cells?

FIG. 5.1

DNA ligation and transformation controls.

the no ligase/no insert transformation control plate even if the gel of ligation products appears as it should, since the sensitivity of transformation is higher. Note that these colonies will not contain insert and should not be selected for screening in later labs.

Keep in mind that any DNA fragments with compatible cohesive ends may be ligated together. This means that in the ligation test tube, anything goes! Molecules may exist with one vector plus one insert (your desired clone), with recircularized vector alone, with one vector and multiple inserts, with vector with incorrect inserts, with concatemerized inserts and with multiple vector combinations (Figure 5.2). Although they do occur during the ligation reaction, it would be unlikely to recover insert-only, or concatamerized insert with no vector from a transformant, because such DNA molecules would be unlikely to contain an origin of replication and an antibiotic resistance gene, and therefore could not replicate *in vivo*. Given the parameters of the protocol, the most likely transformants to be obtained will be either the desired clone or vector alone with no insert. In later exercises, you will learn how to screen for the desired clone.

Laboratory Exercises

Ligations and Ligation Controls

Set up three ligations (or mock-ligations) in microcentrifuge tubes. Label clearly with lab day, station number and sample number. To determine the amount of pET-41a to add, you need to know the concentration of your DNA. Use the equation:

Volume needed (µl) = Amount desired (ng)/Concentration (ng/µl)

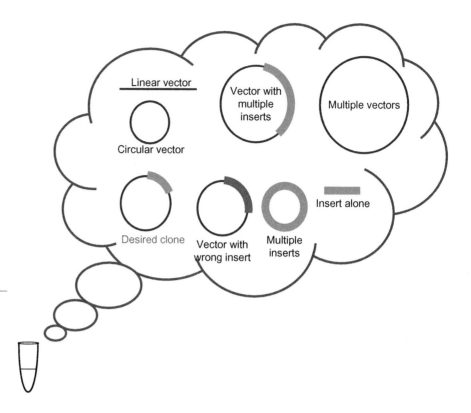

FIG. 5.2

The products of ligation. While many combinations can form in a ligation mixture, after transformation we will most likely detect transformants containing vector only and transformants containing our desired clone (vector plus insert).

Since you need to add 50 ng, use:

$$\text{Volume needed } (\mu l) = 50 \, ng/\text{Concentration } (ng/\mu l)$$

To calculate the amount of water to add, it is 20 μl minus all other volumes.

Ligation (1) 1:3 molar ratio pET-41a: *egfp* insert (experimental):

50 ng pET-41a *NcoI/NotI*	_____ μl
18 ng *egfp* insert	_____ μl
sterile dH$_2$O	_____ μl
10× ligase buffer	2 μl
ligase	1 μl
	20 μl total

Ligation (2) linear pET-41a no ligase control (to control for the presence of uncut vector):

50 ng pET-41a *NcoI/NotI*	_____ μl
sterile dH$_2$O	_____ μl
	20 μl total

Ligation (3) circular (uncut) pET-41a control (to control for viability/competency of cells and your technique):

50 ng pET-41a uncut	_____ μl
sterile dH$_2$O	_____ μl
	20 μl total

Mix gently, and bring contents to the bottom of the tube either by a 5 second centrifugation or by tapping the contents to the bottom of the tube. Incubate for 10 minutes at room temperature. Proceed to transformation.

Note to instructor: This protocol uses NEB (New England Biolabs) T4 DNA ligase, catalog number M0202L. If using a different brand of T4 DNA ligase, follow manufacturer's instructions. The incubation time may be longer.

Divalent Cation-Mediated Transformation

Competent cells are extremely fragile. Never vortex or pipette competent cells vigorously because they will lyse. Never warm competent cells to room temperature because they will start repairing their membranes and will lose their competency. Do not centrifuge the competent cells before use, because you will be unable to get them back into suspension. Set up three transformations (one for each ligation) as described below:

1. Set the heat block to 42°C and fill holes with water. Verify the temperature using a thermometer, as the digital displays are not always accurate.
2. Label three sterile microcentrifuge tubes 1–3 (designations from ligation). Chill and keep on ice.
3. Obtain 75 µl of competent cells, keeping on ice. Keep competent cells on ice at **all** times.
4. "Finger flick" to be sure cells are in suspension, and aliquot 20 µl into each of the three chilled microfuge tubes.
5. Add 2 µl of each of the ligation mixes into the appropriate tube, being sure to add the DNA directly into the bacterial suspension. Mix the tube gently.
6. Incubate on ice for 5 minutes.
7. Heat pulse the cells for 30 seconds at 42°C. The duration and temperature of the heat pulse is critical.
8. Incubate tubes on ice for 2 minutes.
9. Add 80 µl of room-temperature SOC broth or LB broth per tube and shake for 45 minutes at 37°C. (*Start your agarose gel during this downtime.*)
10. Label three LB/kan plates with your initials and 1–3 (for the individual ligations).
11. Mix the cells in the tube gently, then plate all 100 µl from each tube onto LB/kan plates that are labeled 1–3. Each plate should also be labeled with your initials. *Always label the backs of Petri dishes rather than the lids to avoid mixups.*
12. Place the plates inverted (lid down) in the 37°C incubator overnight.

Electrophoresis of Ligation Reactions

This can be started during the 45-minute transformation incubation. While the transformation of the control ligations is a much more sensitive method to detect circularization of the vector (or vector that was not digested to completion), gel electrophoresis of the ligation products will give us more rapid feedback.

Unlike linear DNA, circularized plasmid DNA has a secondary structure, and this secondary structure affects the movement of the DNA through the gel. Therefore, circular plasmid DNA will often run at an unpredictable molecular weight. Additionally, if ligation has occurred, multiple products may be formed, giving rise to multiple bands. In this way, we can see right away whether ligation was successful. If only a single vector band and a single insert band are apparent in the experimental ligation, then it would be a strong indicator that the ligation reaction was unsuccessful.

To prepare your samples for gel electrophoresis, pipette 10 µl of the following samples into five new, labeled microcentrifuge tubes. Then add 2 µl of loading dye to tubes 2–5 before loading into wells.

1. 10 µl 1 kb ladder (premixed with loading dye)
2. 10 µl ligation 1
3. 10 µl ligation 2
4. 10 µl ligation 3
5. 10 µl digested PCR product.

Run the gel as previously described.

When analyzing your gel, if the experimental ligation (lane 2, ligation 1) works, you will typically see the band corresponding to cut vector disappear and bands lining up with uncut vector appearing, as well as some additional bands. If the experimental ligation appears to contain linear vector, it is likely that there is a problem with either your ligase or your ligase buffer. If the vector has cut to completion, lane 3 (ligation 2) should show a single band, corresponding to linear DNA.

Discussion Questions

1. In the cloning procedure we used in the lab, it is possible, albeit unlikely, for multiple inserts to form a concatamer and be cloned into a single expression vector. If three *egfp* inserts were cloned into a single pET-41a expression vector, would EGFP be expressed (following IPTG induction), and if so, would these colonies have more fluorescence compared to clones that have one copy of the *egfp* gene (following IPTG induction)?

2. In our ligation, we used a 1:3 molar ratio of vector to insert. How could different molar ratios (1:1 or 1:10) affect the ligase reaction? What are the pros and cons of excess vector versus excess insert?

3. You perform a ligation with *Nco*I/*Not*I-digested pET-41a vector and the *egfp* insert from pEGFP-N1 (as in the lab). You then transform the ligation mix into *E. coli* and plate on LB medium containing kanamycin. Consider the following unexpected outcomes, and suggest controls:

   a. Nothing grows. What controls might you design to determine whether the *E. coli* cells are viable (alive) versus whether the *E. coli* is competent? What would the results of those controls be?

   b. You see lots of growth, but when you isolate plasmid from numerous *E. coli* colonies, all you find is pET-41a with no insert. Suggest the most likely way these colonies arose (aside from contamination). What would the results of this control look like if your suspicion was correct?

Reference

1   Dugaiczyk A, Boyer HW, Goodman HM. Ligation of *EcoRI* endonuclease-generated DNA fragments into linear and circular structures. *J. Mol. Biol.* 1975;96:171–184.

PART 2

Screening Transformants

The next several experiments are aimed at determining which of the bacterial clones carry the *egfp* gene and can produce the GST::EGFP fusion protein. The monoclonal antibody probe will test for protein expression. The restriction mapping experiment will confirm that an insert of the correct size is present. It is important to make sure that a recombinant molecule to be used for large-scale protein expression comprises exactly what is intended: one copy of the vector and one copy of the insert. The PCR experiment will determine if the gene is present in the correct orientation with respect to the promoter. The visualization of green fluorescence will reveal whether the recombinant protein is folded correctly. Finally, the recombinant plasmid will be sent to a sequencing facility to confirm everything was correctly ligated and that no single nucleotide errors were incorporated during the PCR amplification of the *egfp* gene. If this was your own research instead of a laboratory course, you would likely use only one or two of the screening techniques along with DNA sequencing. It is important to know and understand a wide variety of screening techniques because if problems occur, it is helpful to use alternate techniques for troubleshooting.

Colony Hybridization

The first steps in characterizing the results of a transformation experiment are to determine if any of the clones contain plasmids with insert DNA, whether that DNA is inserted correctly and if the protein can be made.

To screen for the expression of the protein, the probe you use will be a monoclonal antibody to the enhanced green fluorescent protein. While this experiment directly screens for the presence of EGFP protein, we can infer that proper DNA sequence is present in the correct orientation and reading frame if the recombinant protein is made.

This week's lab requires an interim laboratory session because you will need to count your colonies and replica plate colonies at least one day before your regular laboratory period. The experiments you begin during the regular laboratory period this week will be completed over the next two weeks.

Molecular Biology Techniques. DOI: 10.1016/B978-0-12-385544-2.00006-5

LAB SESSION 6A

Interim Laboratory Session

Goal: Today you will count and record the number of transformants on each plate. You will also make replica plates of putative transformants from your experiment. These replica plates will be used to screen for positive clones (those that express the enhanced green fluorescent protein).

Introduction

Today you will prepare the templates to be used in the screening. Replica plating is the technique by which each colony/clone is inoculated onto multiple plates according to a numbered scheme. This method allows each clone to be tested by a variety of methods, while retaining a master plate from which clones can be picked. Although one plate can be used both as the master and as the template for filter hybridizations, the risk of environmental microbial contamination and cross-contamination between colonies is greater. Because positive clones should occur at a high rate, we only need to screen in the order of tens of colonies, so we will pick each colony individually and inoculate replicas using a toothpick. In cases where hundreds or thousands of colonies must be replicated, a sterile velvet stamp may be touched to the original plate and stamped onto multiple blank plates to grow replicas. The individual fibers of the velvet act as tiny inoculating needles.

Laboratory Exercises

Counting Transformants and Replica Plating

Count and record the number of transformants on each of your ligation/transformation plates. You should have many colonies on the plate from the uncut vector control ligation/transformation 3 and several colonies from your experimental ligation 1. You should have very few, if any, colonies on the plates where the bacteria were transformed with ligation 2. Why?

Determine whether your controls gave you the expected results. If they did not, what technical problem could have occurred? Would you still screen the colonies found on plates where you expect to see positive results (vector ligated with insert)? If you see numerous colonies in ligation 2, it may be difficult to find transformants that have incorporated insert in ligation 1. Why?

Replica Plating

We will be doing duplicate replica plates to analyze potential positive clones from your vector–insert ligations.

1. Obtain two plates of LB/kan. One will be used for the antibody probe experiment and the other retained as a master. IMPORTANT: Be sure to label both plates with your lab day and your station number. Always label the bottom of the petri plate, not the lid.

2. Adhere grid stickers to the backs of two fresh LB/kan plates. If grid stickers are not available, create a numbered grid of at least 30 squares on the bottom of your plates with a Sharpie marker.

3. Obtain a plate containing pBIT (positive control) and a plate containing pET-41a (negative control). Using a sterile toothpick, pick up cells from a culture of the positive control (pBIT) by lightly touching a single colony. Lightly drag the toothpick in the shape of a plus in square 19 (or, depending on the grid, toward the middle of the plate). Do not puncture or gouge the agar. Repeat the dragging motion for both plates (do not pick up more cells between plates). Discard the toothpick into a biological waste container, and take a new toothpick. Repeat with pET-41a, but make the streak in the shape of a minus (for negative control) in square 20 for both replicas.

   You are putting the controls toward the center of the plate, because colony hybridizations tend to be most accurate in the center and least accurate at the edges of the plate. It is critical that the controls give a clear result; otherwise none of the data may be analyzed. If you drew your own grids, be sure to plate the controls toward the center of the plate.

4. Pick up cells from your 1:3 vector: insert transformation plate. Lightly drag the toothpick in a diagonal direction in square 1 on the first plate and then, with the same toothpick, in an identical manner on the second plate. To avoid contamination between clones, do not streak all the way to the corners of the grid square because bacteria will spread as they grow. Try not to stab the toothpick into the medium.

5. Using a different autoclaved toothpick for each colony, pick up cells from your plate of transformants. Be sure only to pick well-isolated colonies – not colonies that are touching each other or very close together.

6. Each group should pick 30 of their transformants from ligation 1. If you have fewer than 20 colonies, pick at least 6 colonies from a "mixed unknown" plate. The mixed unknown plate contains a mixture of positive clones and clones with vector only. You may also borrow a plate from another group – you can re-pick colonies they have already picked. Be sure to record in your notes which plate each clone came from. Do not pick colonies from ligations 2 or 3 – these will all be negative.

7. Ordinarily, you would incubate the plates in a 37°C incubator overnight. Instead, stack your plates, wrap colored tape around them (once) and store them, lid down, at 4°C in designated drawers in the refrigerator. Your instructor will transfer them to the 37°C incubators the afternoon prior to your lab session, to ensure that the plates will be warm, fresh and free of condensation for your experiment.

Colony Hybridization: Monoclonal Antibody Probe

Goal: You will screen your clones using an α-EGFP monoclonal antibody probe to determine which of your clones express the enhanced green fluorescent protein. This exercise will be done in two parts: you will prepare the blots this week and probe them in the next lab session.

Introduction

Inserting the *egfp* gene into the pET-41a expression vector in the correct location, orientation and reading frame allows the GST::EGFP fusion protein to be made when cells containing the plasmid are exposed to IPTG. Probing the colonies with a monoclonal antibody probe that binds specifically to the EGFP moiety of the fusion protein will confirm whether the recombinant fusion protein is being correctly translated. From this experiment, one can infer not only that the gene of interest is present, but also that it is in the correct orientation and reading frame.

Nomenclature for antibodies is often represented with the Greek letter "alpha" (α) or the prefix "anti." For example, an antibody that specifically binds to the EGFP protein may be referred to as "anti-EGFP" or "α-EGFP."

The most common method for detecting specific proteins on blots is to hybridize with a primary antibody specific for the protein of interest, followed by incubation with a secondary, tagged antibody that recognizes the first antibody (Figure 6.1).

Before beginning the hybridization, it is critical to induce the expression of the EGFP protein with isopropyl-β-D-thiogalactopyranoside (IPTG) or the protein will not be made. After protein expression, the bacterial cells must be lysed. This is done using non-denaturing conditions that preserve protein structure so that it can still be recognized by antibodies in the subsequent steps. Following cell lysis, the proteins are transferred to a nitrocellulose membrane, which naturally binds protein.

Once the colonies are lifted, the blots will be incubated in a solution with large amounts of non-specific protein (bovine serum albumin or powdered nonfat milk) to block non-specific antibody binding directly to the nitrocellulose membrane.

Next, the blot is probed with the primary antibody. Our primary antibody is an α-GFP monoclonal antibody (mAB) that recognizes and binds to a conserved region on the closely related EGFP protein. Unbound and non-specifically bound antibody is then washed off with a buffer containing a

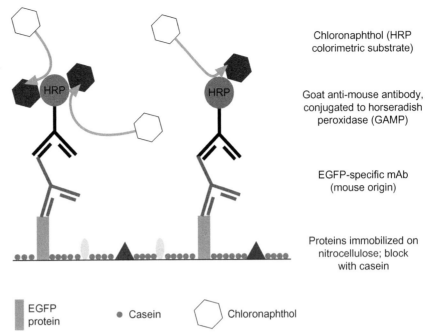

FIG. 6.1

The principle of monoclonal antibody detection. (1) Cellular proteins bind to the nitrocellulose membrane. (2) The membrane is then blocked with a protein that will not interact with the primary antibody, such as casein found in nonfat milk. (3) Primary antibody is added and allowed to bind to the specific epitope on the EGFP protein and unbound antibody is washed away. (4) Secondary antibody conjugated to horseradish peroxidase (GAMP) is added and allowed to bind to the primary antibody and then excess (unbound) antibody is washed away. (5) Finally, the colorimetric substrate chloronaphthol is added. Spots from transformants that expressed the EGFP fusion protein will appear purple; clones that did not express the protein will remain colorless.

Chloronaphthol (HRP colorimetric substrate)

Goat anti-mouse antibody, conjugated to horseradish peroxidase (GAMP)

EGFP-specific mAb (mouse origin)

Proteins immobilized on nitrocellulose; block with casein

EGFP protein   ● Casein   Chloronaphthol

nonionic detergent such as Tween 20 at as low a concentration as possible to preserve epitope integrity.

The blot is then probed with the secondary antibody. The secondary antibody recognizes the species-specific, conserved portions of the primary antibody on one end and is conjugated with the enzyme horseradish peroxidase on the other end. Because the primary antibody is of mouse origin, we will use a secondary antibody made in goats that binds conserved portions of mouse antibodies. We will refer to the secondary antibody as GAMP (**G**oat **A**nti-**M**ouse antibody conjugated to horseradish **P**eroxidase). The horseradish peroxidase bound to the secondary antibody allows for its detection.

Finally, detection of the protein of interest is achieved by incubating the blots in the colorimetric substrate chloronaphthol. Horseradish peroxidase, conjugated to the goat anti-mouse secondary antibody, cleaves chloronapthol to make a purple precipitate. A flowchart of the monoclonal antibody probe procedure is outlined in Figure 6.2.

Laboratory Exercises

Colony Hybridization with an α-EGFP Monoclonal Antibody Probe: Part 1

IMPORTANT: When working with membranes, always wear gloves. Also, label blots in PENCIL (do not use ink).

1. Prepare a nitrocellulose membrane for probing with α-EGFP. Use scissors to make two asymmetrical small notches in a nitrocellulose membrane if they are not already present. This will be used to orient your membrane with the original plate. Label the matte side (if your membrane is only coated on one side) of the membrane with your lab day

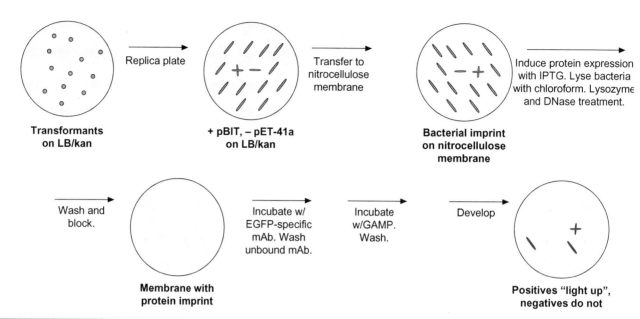

FIG. 6.2

Flowchart of the colony hybridization with an
α-EGFP monoclonal antibody probe experiment.

and station number. *In later steps, it is important that the matte side of the membrane is the side that comes in contact with the bacteria.*

CAUTION: Never directly touch nitrocellulose membranes. Oils from your skin will inhibit the transfer of solutions to the membranes and proteins from your finger tips will bind irreversibly. Always wear gloves and manipulate the membranes with blunt-ended forceps.

2. Lay the notched nitrocellulose membrane in 750 μl of a solution of IPTG (20 mg/ml) on a piece of plastic wrap. The IPTG will induce synthesis of the fusion protein.
3. Place the membrane on a replica plate of your transformants (labeled side toward the colonies). To prevent air bubbles, carefully drop the membrane near one edge of the plate and lightly roll the membrane across the rest of the plate. Do not slide or reposition the membrane or the colonies will smear. Mark the bottom of the Petri plate with a Sharpie marker to show the precise location of the notches. Label the plate mAB.
4. Pick up the membrane and "flip" it over onto a fresh LB/kan plate so that the bacterial colonies are on top of the membrane (facing up).
5. Label the plate and place the membrane and LB/kan plate in a 37°C incubator for 2 hours.
6. Lyse the bacterial cells by exposing them to chloroform vapors for 30 minutes in the fume hood. Remove the top Petri plate and invert the membrane and bottom plate over a pool of chloroform. Glass Petri plates with support mesh will be available in the chemical fume hood; pour a small amount (5–10 ml) into the glass plate, cover with mesh, and invert the bottom of the LB/kan plate with your membrane directly over the chloroform. Do not remove the membrane from the LB plate; allow it to remain attached to the agar during the time required for lysis.

CAUTION: Chloroform will disintegrate plastic Petri plates on direct contact; make sure that the chloroform aliquoted for bacterial lysis is in a glass container and that a support mesh or tape is used to prevent direct contact with the inverted plate. Chloroform should be aliquoted using a glass pipette or Erlenmeyer flask and should be used only in the fume hood.

7. Remove your membrane from the hood. Separate the membrane from the agar and wash the membrane, colony side up, for 30 minutes in a Petri plate containing 7.5 ml of lysis buffer [blocking solution with lysozyme (40 μg/ml) and pancreatic DNase (1 μg/ml)] to remove bacterial debris.

8. Incubate the membrane in 10 ml of blocking solution without lysozyme or DNase for 15 minutes. Use a platform shaker to gently swirl the solution.

9. Rinse in 1× wash buffer with IGEPAL (a mild detergent) plus 0.1% nonfat powdered milk for 15 minutes at room temperature.

10. Rinse in 1× wash buffer with 0.1% nonfat powdered milk for 10 minutes.

11. Place blot on Whatman paper and allow it to dry for 5 minutes. Sandwich the blot between two squares of fresh Whatman paper and give it to your laboratory instructor. The blots will be frozen and we will continue with this protocol in the next lab session.

Discussion Questions

1. Is it possible to receive a false positive in this experiment: could spots on the membrane appear purple even if the corresponding clone is not producing EGFP? Is it possible to receive a false negative in this experiment? If so, what are the possible causes?

2. What outcome might you predict if you neglected to block the membrane with casein prior to the addition of the primary antibody?

3. You want to perform a colony hybridization using a primary antibody that was raised in pigs. Unfortunately, you can't find any secondary antibodies in the refrigerator in your lab, so you are going to have to buy some before performing the experiment.
   a. List the critical features that the secondary antibody "must" have.
   b. Using the catalog for the antibody supplier www.abcam.com, give the name and catalog number of a suitable secondary antibody you could buy and use for your experiment.

LAB SESSION 7

Characterization of Recombinant Clones: Part 1

In this laboratory session, you will complete the colony hybridization with the monoclonal antibody probe experiment; you will also begin a PCR screen.

Molecular Biology Techniques. DOI: 10.1016/B978-0-12-385544-2.00007-7

LAB SESSION 7A

Completion of Colony Hybridization with a Monoclonal Antibody Probe

Goal: In this exercise, you will complete the colony hybridization with an EGFP-specific monoclonal antibody probe and analyze the results.

Introduction

Review the Introduction from Lab Session 6B.

Laboratory Exercise

Colony Hybridization with an α-EGFP Monoclonal Antibody Probe: Part 2

1. Retrieve your blot from the freezer and rinse in 1× wash buffer with 0.1% nonfat powdered milk for 5 minutes in a clean Petri dish at room temperature.
2. To prepare the primary antibody, dilute the mouse α-EGFP 1:1000 by adding 7.5 μl α-EGFP to 7.5 ml blocking solution.
3. Pour the antibody solution over your blot and incubate the plate at room temperature for 1 hour on a platform shaker at 40 rpm.
   (Proceed to Lab Session 7B during this incubation period.)
4. Wash the membranes in each of the following buffers. After each wash, discard the wash solution in the sink, then invert the Petri plate and drain the remaining wash solution on paper towels. Then add 10 ml of the next solution.
   - wash 1 (10 ml): 1× wash buffer + 0.1% nonfat powdered milk (5 minutes);
   - wash 2 (10 ml): 1× wash buffer + 0.1% nonfat powdered milk + 0.1% IGEPAL* (10 minutes);
   - wash 3 (10 ml): 1× wash buffer + 0.1% nonfat powdered milk (10 minutes).
5. Add 7.5 ml of the secondary antibody, goat anti-mouse antibody conjugated to horseradish peroxidase (GAMP), diluted 1:500 in antibody blocking solution (15 μl of GAMP plus 7.5 ml of blocking solution). Incubate the plate for 30 minutes on a platform shaker at 40 rpm, at room temperature.

*IGEPAL is a nonionic, non-denaturing detergent that is also known by the following names: Nonidet P40; NP 40; IGEPAL CA-630 and Nonylphenyl-polyethylene glycol.

6. Detergents inhibit peroxidase activity. Therefore IGEPAL is omitted in the following steps. Wash the plate three times with the following:
   - wash 1 (10 ml): 1× wash buffer + 0.1% nonfat powdered milk (5 minutes)
   - wash 2 (10 ml): 1× wash buffer + 0.1% nonfat powdered milk (5 minutes)
   - wash 3 (10 ml): 1× wash buffer + 0.1% nonfat powdered milk (10 minutes).
7. After the last wash, incubate the membrane in 7.5 ml of peroxide stain for 15–20 minutes at room temperature. A purple color should develop.
8. Wash the membrane in two changes of distilled water.
9. Dry the membrane on paper towels. Keep in the dark to maintain color.
10. Analyze your results by comparing to your original replica plate. Record which positives correspond to which transformant number and include the information in your laboratory notebook.

LAB SESSION 7B

PCR Screening

Goal: This PCR screen will confirm whether clones have the *egfp* insert and whether it is in the correct orientation in the vector. You will begin this experiment today and analyze the results next week.

Introduction

Because we inserted the *egfp* gene using the method of forced cloning, we are confident that the gene could only have been inserted in the correct orientation. This screening technique, however, is also able to confirm the correct orientation of an insert in cases where that is an issue (when forced cloning is not used and cloning is performed with only one enzyme).

The key to this experiment is where the PCR primers are designed to anneal. As in any PCR reaction, the two primers flank a DNA region, with one primer annealing to the upper strand and one to the lower strand, with the primers facing inwards toward each other. The special feature of these primers is that one primer must anneal to the insert sequence and one must anneal to the vector sequence. If the insert is present and in the correct orientation, a PCR product that is the size of the flanked DNA sequence will be produced (in our case, this is 1183 bp). The product can be visualized by running a small amount of the PCR reaction on an agarose gel. If the insert is not present in the vector, the primer that was designed to anneal to the insert will be unable to bind and no PCR product will be made. If the insert was cloned into the vector in the incorrect orientation, both primers will anneal to the same strand of DNA in the same orientation and exponential amplification will be unable to occur, so no PCR product will appear on an agarose gel. We can therefore ascertain that if a PCR product is visible on an agarose gel, then the insert is present and in the correct orientation (Figure 7.1). Remember that DNA synthesis can only occur in a 5′ to 3′ direction.

Laboratory Exercise

Polymerase Chain Reaction Screen for Recombinant Clones

You will complete this protocol during your 1 hour incubation with mono-clonal antibody from the antibody probe protocol.

The PCR procedure you will follow is similar to that for amplifying isolated DNA except that the first step, 95°C, takes longer in the first cycle because we have not only to denature the DNA, but also to ensure that the bacteria

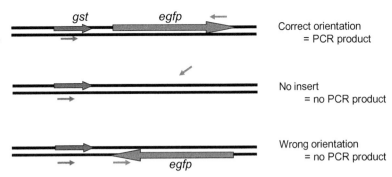

FIG. 7.1

Screening transformants by PCR. The forward primer anneals to the *gst* gene on the vector; the reverse primer anneals to the *egfp* insert sequence. In the first example, *egfp* is inserted in the correct orientation in the vector. The primers anneal in the correct positions and an 1183 base pair PCR product is made. In the second example, *egfp* is not present, so the reverse primer cannot anneal and no PCR product is made. In the final scenario, *egfp* is present in the reverse orientation. Both primers bind, but in the same orientation, to the same strand. No PCR product is made.

→ Forward primer anneals to the vector sequence (*gst*).
← Reverse primer anneals to the sequence in *egfp*.

are lysed by the heat to release the DNA. At 95°C, most cellular enzymes that degrade DNA become heat inactivated allowing PCR to occur without a clean DNA preparation.

Each station will receive one tube of master mix, or you may be asked to make your own master mix following the chart below. Using a master mix and then aliquotting from the mix both saves time and reduces variability between reactions.

1. The recipe for the master mix is as follows (students will do eight reactions, 20 μl per tube):

Item	per 180 μl
Water	155.2 μl
dATP	0.36 μl
dTTP	0.36 μl
dCTP	0.36 μl
dGTP	0.36 μl
Primer 1 (pBITrev)*	0.9 μl
Primer 2 (pBITfor)*	0.9 μl
10× Buffer with Mg	18 μl
Taq polymerase	3.6 μl

*The pBITrev primer anneals to the 3′ portion of the sense strand of the *egfp* gene. The sequence of the pBITrev primer is CTTGTACAGCTCGTCCATGC. The pBITfor primer anneals to the vector sequence (specifically, the sequence in the middle of the antisense strand of the *gst* gene). The sequence of the pBITfor primer is CAAGCTACCTGAAATGCTGA. The predicted PCR product is 1183 base pairs in size.

2. Aliquot 20 μl of reaction mix into eight small strip tubes. You will screen five putative positive transformants, and include a no-template DNA control to test for contaminant DNA, a pBIT positive control, and a pET-41a negative control (eight in total).

3. Use a sterile yellow pipette tip (wooden toothpicks contain an inhibitor of PCR) to pick up a small amount of bacterial growth from a putative transformant from your freshest master plate. Swirl the tip in PCR reaction mix. Label the PCR tube with the number on your master plate. Repeat with a different clone for each tube (including controls).

55

For this experiment, LESS IS MORE. Pick up a very small amount of bacterial growth, about the amount that would fit on the head of a pin. Putting too much bacteria in the PCR reaction will result in degradation of the PCR product.

4. Run the following program:
   Step 1: 95°C, 10 minutes (denature).
   Step 2: 95°C, 1 minute.
   Step 3: 56°C, 1 minute.
   Step 4: 72°C, 1.5 minutes.
   Step 5: Repeat steps 2–4 twenty-nine times.
   Step 6: 72°C, 5 minutes to extend all unfinished products.
   End.

You will analyze the results next week using gel electrophoresis.

LAB SESSION 7C

Prepare Fresh Replica Plate

The bacteria on your replica plate will lose some viability over time. Prepare a fresh replica plate from your old master replica plate. Be sure to label your initials and today's date so you will know it is your freshest plate. Incubate inverted at 37°C overnight; your instructor will move the plates to the refrigerator for storage.

Discussion Questions

1. In our PCR screen, what would you hypothesize the result would be if an incorrect insert was cloned into the vector and:
   a. the incorrect insert was a different size than the correct one?
   b. the incorrect insert was the same size as the correct one?
2. When performing the PCR screen, why could putting too much bacteria in the PCR reaction mix cause possible degradation of the PCR product?
3. What would happen if you used primers in the PCR screen that bind to vector DNA on opposite sides of the insert (neither primer designed to anneal to the insert)? What conclusions could you draw if the experiment were designed this way?

57

LAB SESSION 8

Characterization of Recombinant Clones: Part 2

This week you will analyze the results of the PCR screen that was started last week. You will also isolate plasmid DNA from transformants on which you will subsequently perform restriction digestion analyses next week. Finally, you will begin a fluorescence visualization assay.

Molecular Biology Techniques. DOI: 10.1016/B978-0-12-385544-2.00008-9

LAB SESSION 8A

Interim Laboratory Session

Goal: Inoculate cultures to be used to isolate plasmid DNA for restriction analysis during your next laboratory session.

Laboratory Exercise

Inoculate Cultures for Minipreps

1. Select five out of the six transformants you performed the PCR screen on.
2. Label six culture tubes (containing 2 ml LB/kan) with your lab day and station number. Label five of the tubes with the selected transformant numbers and label one tube as pBIT (positive control).
3. Heavily inoculate each 2 ml LB/kan culture with the putative trans-formant or positive control as on the label. Inoculate tubes by picking bacterial growth off the Petri dish using a sterile toothpick and then dropping and leaving the toothpick in the tube containing growth medium, being careful only to touch the end of the toothpick that is not in contact with the medium.

   This inoculation method does not utilize the principle of aseptic technique. It typically works well for inoculating cultures for minipreps because the antibiotics in the medium keep most contaminants from growing and because the inoculum is so large that it would likely out-compete any contaminant. In cases where a pure culture is critical, such as creating a freezer stock, do not utilize this method.
4. Place your tubes in the appropriately labeled test tube rack at the front of the lab. Your instructor will place these in the refrigerator at 4°C until the afternoon prior to your lab session and will then incubate them at 37°C overnight so cultures will be fresh for your lab session.
5. Cleanup: Discard all old plates except for the most recent replica plate and the IPTG plate. Use biohazard bags for disposal.

LAB SESSION 8B

Analysis of PCR Screen Results

Goal: You will analyze the products of the PCR screening technique performed last week by agarose gel electrophoresis. Recall that only reactions from clones containing the *egfp* insert will produce an 1183 bp PCR product. Clones that do not have the *egfp* insert will produce no PCR product.

Introduction

Review Lab Session 7B.

Laboratory Exercise

Gel Electrophoresis and Analysis of PCR Samples from Last Week

1. Add 2 μl of loading dye to each PCR reaction. Mix gently.
2. Pour a 1% agarose gel containing GelRed, as described in Lab Session 3.
3. Load the gel in the following order:
   - Lane 1: 10 μl NEB 1 kb ladder (premixed with loading dye);
   - Lanes 2–8: 15 μl of PCR reaction from Lab Session 7B per lane (with loading dye). If you are using an eight-well gel, leave out the no template PCR control from the gel.
4. Run the gel until the tracking dye runs halfway through the gel.
5. View the gel under UV light and photograph. Positive clones will have a PCR product of approximately 1.2 kb, and negative clones will have no PCR product. Record your results in your lab notebook.

Isolation of Miniprep DNA from Potential Transformants

Goal: You will isolate the plasmid DNA from several transformants and then digest the plasmids with the restriction endonucleases *Nco*I and *Not*I to confirm the presence of the *egfp* gene.

Clones containing the vector with *egfp* successfully inserted will contain two visible fragments on the gel following digestion: 724 bp (corresponding to *egfp*) and 5859 bp (corresponding to the vector). Clones containing vector only will show one visible fragment on the gel of 5859 bp (corresponding to the large vector fragment), although another fragment too small to visualize on the gel is also present (73 bp, corresponding to the small fragment of DNA between *Nco*I and *Not*I on the vector).

This experiment will be carried out over the course of two weeks. This week you will purify the plasmid DNA and next week perform the restriction digestion and analysis. The plasmid DNA obtained from this miniprep will also be used for DNA sequencing.

Introduction

Regardless of initial methods used to identify positive transformants, you will need to isolate DNA for the final analysis. We will use the same protocol we used at the beginning of the semester to purify plasmid DNA.

Remember, unexpected things can happen in a ligation: small amounts of stray DNA, as well as the insert you want, can theoretically be cloned. The bottom line in analyzing transformants is if it matches the pattern you expect, keep it. If it doesn't, throw it out. When analyzing restriction digestion patterns, keep the following principles in mind:

- The DNA migrates at a rate inversely proportional to the size of the DNA fragment. From previous experience, you know that including a lane of standard molecular weight markers is critical for analysis of your DNA. Because the molecular weight of each of the bands in the DNA standard is known, you can use these molecular weight markers to estimate the size of unknown sample fragments.
- Circular supercoiled DNA travels at an unpredictable rate through the gel due to the secondary structure. Therefore, the size of uncut DNA cannot be estimated using a typical DNA ladder. Multiple bands may also occur due to variations in the secondary structure. Whenever you electrophorese restriction-digested plasmid DNA, one lane should contain uncut DNA as a reference.

- The sum of the molecular weight of the fragments generated by digestion of a circular molecule equals the molecular weight of the uncut molecule. (To visualize this, think of cutting a rubber band into pieces; the total length of the pieces is the length of the rubber band.)
- The intensity of band fluorescence is directly proportional to the mass of the fragment. Deviations from this principle are indicative of partial digests or a doublet (two bands of the same size).

Laboratory Exercise

Isolation of Miniprep DNA from Potential Transformants

Miniprep DNA from five of your transformants (four putative positive and one putative negative) and your pBIT positive control. The protocol below is from QIAGEN QIAprep® Miniprep Handbook (Second edition, 12/2006) (http://www1.qiagen.com/literature/render.aspx?id=370).

Plasmid DNA Purification Using the QIAprep Spin Miniprep Kit and a Microcentrifuge

Note: All protocol steps should be carried out at room temperature.

1. Label six microcentrifuge tubes according to the transformants you inoculated earlier in the week, plus the one tube for the positive control. Transfer 1.5 ml of each culture into the correspondingly labeled microcentrifuge tube.
2. Harvest the bacterial cells by centrifugation at 12,000 rpm for 30 seconds. Remove all traces of supernatant by micropipetting.
3. Re-suspend pelleted bacterial cells in 250 μl Buffer P1 and transfer to a microcentrifuge tube. Ensure that RNase A has been added to Buffer P1. No cell clumps should be visible after re-suspension of the pellet. The bacteria should be re-suspended completely by vortexing or pipetting up and down until no cell clumps remain.
4. Add 250 μl Buffer P2 and mix thoroughly by inverting the tube four to six times. Mix gently by inverting the tube. Do not vortex, as this will result in shearing of genomic DNA. If necessary, continue inverting the tube until the solution becomes viscous and slightly clear. Do not allow the lysis reaction to proceed for more than 5 minutes.
5. Add 350 μl Buffer N3 and mix immediately and thoroughly by inverting the tube four to six times. To avoid localized precipitation, mix the solution thoroughly immediately after addition of Buffer N3. The solution should become cloudy.
6. Centrifuge for 10 minutes at 13,000 rpm (17,900 × g) in a table-top microcentrifuge. A white pellet will form.
7. Apply the supernatants from step 6 to the QIAprep spin column by decanting or pipetting.
8. Centrifuge for 30–60 seconds. Discard the flow-through.
9. Wash QIAprep spin column by adding 0.75 ml Buffer PE and centrifuging for 30–60 seconds.
10. Discard the flow-through, and centrifuge for an additional 1 minute to remove residual wash buffer.

IMPORTANT: Residual wash buffer will not be completely removed unless the flow-through is discarded before this additional centrifugation. Residual ethanol from Buffer PE may inhibit subsequent enzymatic reactions.

11. Place the QIAprep column in a clean, labeled 1.5 ml microcentrifuge tube **that has had the lid cut off**. To elute DNA, add 50 µl Buffer EB (10 mM Tris·Cl, pH 8.5) to the center of each QIAprep spin column, let stand for 1 minute, then centrifuge for 1 minute.

12. Label six fresh microcentrifuge tubes with your lab station, date and transformant number. Transfer DNA to these tubes and save in your freezer box until the next lab session.

LAB SESSION 8D

Visualization of Green Fluorescent Protein: Part 1

Goal: You will replica plate your transformants (and positive and negative controls) onto an LB medium that contains kanamycin and IPTG. This will allow for visualization of green fluorescence in positive clones. You will also replica plate onto an LB/kan medium so that you will have a fresh master plate.

Introduction

As discussed in Lab Session 2, IPTG acts to derepress the T7/lac promoter on expression vector pET-41a. By incorporating IPTG into the growth medium, clones that can express the enhanced green fluorescent protein will appear slightly green even under ambient light. Positive clones will fluoresce bright green when UV light is shone on them.

Laboratory Exercise

Green Fluorescence Assay and Preparation of a Fresh Master Plate

You will use your freshest master plate to replica plate onto the LB/kan/IPTG plate, as well as onto a new LB/kan plate.

1. Retrieve your master plate from the refrigerator.
2. Obtain one LB/kan/IPTG plate and one LB/kan plate. Adhere grid stickers to the backs and label with your lab day, station number and today's date. *Always label the backs of Petri dishes rather than the lids to avoid mix-ups.*
3. Using a toothpick, replica plate from your master plate onto the two fresh plates. Do this as you did in Lab Session 6A, using a sterile toothpick to pick a small amount of bacterial growth from each grid square on the master plate and inoculating the same squares on the two fresh plates. Be sure to change toothpicks between each clone you pick, but inoculate both fresh plates with the same toothpick (only pick up the transformant from the master plate once).
4. Tape plates shut and place in an inverted position in the 37°C incubator overnight. Your instructor will move these plates to the refrigerator tomorrow, and you will observe the IPTG plate next week in the lab. The LB/kan plate will be used as a new master plate.

Discussion Questions

1. If the *egfp* gene is approximately 700 bp in length, why do we predict the PCR product from positive clones in Lab Session 8B to be approximately 1200 bp in length?
2. Is it possible for a clone to have multiple copies of *egfp* inserted? Could two copies be inserted? Could three copies be inserted? If so, what would be the result of the PCR screen?
3. If multiple copies of the *egfp* gene could be inserted, would it have more, less, or the same amount of fluorescence compared to a clone with a single copy of the *egfp* gene present?

LAB SESSION 9

Characterization of Recombinant Clones: Part 3

This week you will perform restriction digestion analyses on the plasmid DNA you prepared last time in the lab, as well as complete the fluorescence visualization assay.

Molecular Biology Techniques. DOI: 10.1016/B978-0-12-385544-2.00009-0

LAB SESSION 9A

Characterization of Miniprep DNA from Potential Transformants (Restriction Enzyme Analysis of Putative Transformants)

Introduction

Review Lab Session 8C.

Laboratory Exercise

Restriction Enzyme Analysis of Miniprep DNA

You will digest DNA from pET-41a/*egfp* transformants in order to confirm the presence of the insert. Since we used the *Nco*I and *Not*I restriction sites to clone the *egfp* gene into pET-41a, we will use these enzymes to release the insert. Follow the procedure described below to set up six double digests with *Nco*I and *Not*I. It will be useful to draw a mock gel predicting the number and size of DNA fragments you would expect from the restriction enzyme digestion.

When performing multiple digests, it is helpful to make a master mix containing common reagents found in each reaction. In this case, the master mix is comprised of water, restriction buffer, BSA and the two restriction enzymes (*Nco*I and *Not*I). Once this master mix is made, it is aliquoted to separate tubes to which the unique DNA samples will be added for digestion. Creating a master mix provides a number of distinct advantages compared to pipetting all the reagents into separate tubes. First, since you will be pipetting larger volumes, you will be minimizing the error that is inherent in pipetting smaller volumes. Second, using a master mix reduces variability between samples. You can imagine that if you're attempting to pipette 1 µl of *Nco*I into each of six tubes, some tubes might have 0.9 µl while others might have 1.1 µl. A master mix reduces this variation. Finally, using a master mix is a time saver. You will be pipetting fewer times than if you were to pipette each component into each separate tube for the digestion. To set up your master mix for your digestion reactions, follow the instructions below.

PREPARING THE MASTER MIX

You will need to set up a master mix for the double digests.

1. Determine the reagent volumes (enzyme, restriction buffers, water) for one enzyme digest (see Table 9.1).

Table 9.1 Master mix for double digests

Component	1 tube	7 tubes
Sterile water	5.8 µl	
Restriction enzyme Buffer 3	2 µl	
BSA	0.2 µl	
*Nco*I	1 µl	
*Not*I	1 µl	
DNA	10 µl*	–
Total volume:	20 µl	

*To be added to numbered tubes separately (not in master mix).

Ten microliters of miniprep DNA will usually have between 500 and 1000 ng of DNA. Since most enzymes are sold in concentrations of about 10 units/µl, 1 µl of enzyme is more than enough for one miniprep DNA digestion.

2. Multiply the volumes by the number of tubes plus one, to correct for pipetting errors.

3. Combine first water, then restriction buffer and BSA, and finally enzyme(s) in one tube (master mix). Vortex your tube and keep on ice.

4. Aliquot 10 µl of master mix to each tube. Make sure the tubes are labeled and then add 10 µl of miniprep DNA to each appropriately labeled tube.

5. Incubate the reactions for 45 minutes at 37°C. (You should have six tubes.)

6. Add 1/10 vol (2 µl) loading buffer directly to the tubes.

7. Pour a 1% agarose gel.

8. Load samples in the following order:
   - 10 µl 1 kb ladder
   - 10 µl uncut pBIT (1 µl uncut pBIT DNA + 8 µl water + 1 µl loading buffer)
   - 15 µl digested miniprep DNA 1
   - 15 µl digested miniprep DNA 2
   - 15 µl digested miniprep DNA 3
   - 15 µl digested miniprep DNA 4
   - 15 µl digested miniprep DNA 5
   - 15 µl digested miniprep DNA 68.

Run the gel until the dye-front reaches halfway down the gel,

Negative clones corresponding to the vector without an insert should give a single band the size of the pET-41a vector, approximately 6 kb. Positive clones, which are vector plus *egfp* insert, should produce two bands: approximately 6 kb and 724 bp.

Visualization of Green Fluorescent Protein: Part 2

Goal: You will observe the transformants that you plated on IPTG medium using ultraviolet (UV) light. Positive colonies that contain the *egfp* gene (and therefore are expressing the GST::EGFP fusion protein) will glow bright green!

Introduction

The enhanced green fluorescent protein has an excitation peak at 488 nm (blue light) and emits light maximally at 507 nm. When you shine UV light on your positive transformants, although you are not exciting EGFP at the excitation peak, there is still enough excitation that occurs to allow the colonies to appear bright green. This assay, since it is so easy to perform, will be used to judge the success of the other screening techniques you used in Lab Sessions 6–8. Most proteins are not fluorescent. In a research lab, it is rare to be so lucky as to be trying to clone a gene that has such a simple assay for expression. For our purposes, it serves as an unambiguous positive control that can be used to determine which of the other techniques were the most reliable.

Laboratory Exercise

Visualization of Clones Expressing the Enhanced Green Fluorescent Protein on IPTG Plates

1. Obtain the IPTG replica plate you inoculated last week.
2. Remove the lid and invert the plate (open side down) on a UV transilluminator. *While viewing your transformants on the ultraviolet light box, you must wear a UV protective face-shield, and also protect any skin from direct exposure to the UV light by wearing gloves and/or a lab coat.*
3. Turn on the ultraviolet light and view the plate. The positive control and positive clones will fluoresce bright green due to the expression of the enhanced green fluorescent protein, as shown in Figure 9.1. Record which clones are positive and which are negative in your lab notebook.

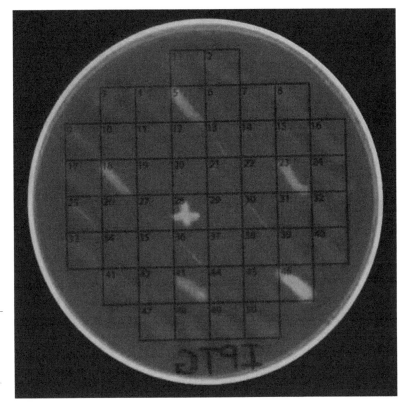

FIG. 9.1

Fluorescence of *E. coli* expressing the enhanced green fluorescent protein. The positive control, pBIT, is inoculated in grid number 28. Clones numbered 5, 18, 23, 43 and 46 are also positive. The rest of the clones are negative.

LAB SESSION 9C

Computational Analysis of DNA Sequence from a Positive Clone: Part 1

Goal: You will verify the DNA sequence of the *egfp* insert you cloned into pET-41a. This experiment will be done over the course of two weeks. First, you will prepare samples for sequencing. These samples will then be sent to a DNA sequencing facility. Second, the data obtained from the sequencing runs will be analyzed to determine the exact sequence of the *egfp* insert and compare it to the known and expected sequence.

Introduction

Cloning by PCR is an effective and straightforward method of cloning a gene into a vector. As discussed in Lab Session 3, cloning by PCR has many advantages over traditional cloning. Despite this, there is one main disadvantage of PCR cloning: errors incorporated during PCR. A single base mutation in your PCR product can lead to amino acid changes or early termination of your fusion protein during translation. While high-fidelity DNA polymerases, such as Vent, can be used to minimize these errors, mutations can still occur. To ensure that subsequent experiments are performed with your expected protein and not a mutated version, verifying the DNA sequence of your plasmid is imperative.

DNA sequencing has become a routine practice in molecular biology and is used in numerous applications. DNA sequencing was initially performed in the 1970s, with labor-intensive, expensive techniques. This technology evolved, largely due to the work of Frederick Sanger, who received the second of his two Nobel Prizes in Chemistry for his work in developing the technology of DNA sequencing.[1] This chain-termination method of DNA sequencing, also known as Sanger sequencing, is what you will use to analyze the DNA sequence of a positive transformant. Other DNA sequencing platforms are used for high-throughput applications such as whole-genome sequencing. These methodologies can obtain up to hundreds of millions of base pairs of sequence in a day. Most forms of high-throughput sequencing are still more expensive per run than the traditional chain-termination method. Furthermore, to sequence your *egfp* insert, you only need to sequence about 750 nucleotides, not millions.

A typical sequencing reaction contains DNA template, DNA polymerase, one sequencing primer, deoxynucleotide triphosphates (dNTPs) and dideoxynucleotide triphosphates (ddNTPs). The key players in DNA sequencing by chain termination are the ddNTPs. As shown in Figure 9.2, the DNA to be sequenced serves as a template for synthesis for complementary

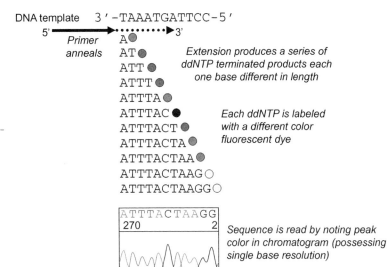

FIG. 9.2

DNA sequencing using fluoresent ddNTPs, also known as Sanger sequencing. Fluorescently labeled ddNTPs act as chain terminators during the synthesis of complementary DNA fragments. These fragments are separated using capillary electrophoresis and results are obtained in the form of a chromatogram. Reprinted from *Fundamentals of Forensic DNA*, JM Butler, page 65, Copyright (2009)[2].

strands of DNA. The sequencing primer anneals to the template, and DNA polymerase binds to the template/primer hybrid DNA and synthesizes a new DNA strand using the nucleotides present in the reaction mixture. The dNTPs are the "normal" bases (A, C, G and T) and can be added to each other, one after another, via phosphodiester bonds. Conversely, the ddNTPs lack the 3′ hydroxyl group that is necessary to form a phosphodiester bond. Because of this, if a ddNTP is incorporated into the DNA strand during synthesis, no additional synthesis can occur. In other words, it acts as a chain terminator. The ddNTPs are present in the sequencing reaction in limiting quantities. Thus, the majority of the time, a dNTP is added and synthesis continues; only occasionally does a ddNTP get added. Because the addition of a ddNTP is random, the result of a sequencing reaction is partial copies of the DNA template, varying in size. Each fragment will start at the 5′ end with a sequence corresponding to the primer sequence and will terminate wherever a ddNTP was added.

The determination of the DNA sequence is dependent on reading the lengths of these fragments and knowing which ddNTP occupies the 3′ position. This is most typically done by using fluorescently-labeled ddNTPs (each ddNTP is labeled with a different color) in conjunction with capillary electrophoresis in automated DNA sequencers. Each sequencing reaction contains a pool of synthesized DNA fragments with different lengths, and each fragment will fluoresce according to which ddNTP was incorporated at its 3′ end. With automated DNA sequencing, these fragments are separated by size using capillary gel electrophoresis, with the smallest fragments traveling through the matrix fastest. As the fragments exit the capillary column, a laser excites the sample and a computer records the fluorescence emitted from the sample. The resulting data, called a chromatogram or electropherogram, shows a series of peaks which reflect the DNA sequence of a synthesized strand of DNA. **An excellent visual representation of DNA sequencing by chain elongation can be found here: http://www.dnalc.org/resources/animations/cycseq.html**.[3]

In today's lab session, you will assemble sequencing reactions to send to a DNA sequencing facility. When performing DNA sequencing, the selection of your sequencing primers is one of the most critical aspects of the experiment. Typically, a successful sequencing run yields 800–1000 nucleotides of sequence. Since the *egfp* sequence is 720 nucleotides in length, we should be able to sequence the entire gene using a primer that anneals just upstream of the beginning of the *egfp* sequence (Figure 9.3). It is worth noting that the interpretable sequence will start 15–30 nucleotides downstream of the primer, so never design sequencing primers to the very beginning of the region you are trying to sequence.

While we are able to sequence the entire *egfp* sequence using one primer, there are many times when using additional primers is necessary. If your insert sequence is over 1000 nucleotides in length, you would need multiple primers to cover the entire sequence. These primers can be designed to either strand of the template; often researchers will use antiparallel primers flanking a gene and sequence the DNA in both directions. Also, if there are uncertainties or discrepancies in your sequencing results, performing an additional sequencing reaction using a different primer (i.e. one that binds to a different region of the DNA) can clarify the sequence. Finally, if you are sequencing DNA with an unknown sequence (not confirming a known sequence as we are doing in this exercise) always use multiple primers resulting in overlapping sequence results to most accurately determine the sequence of your DNA.

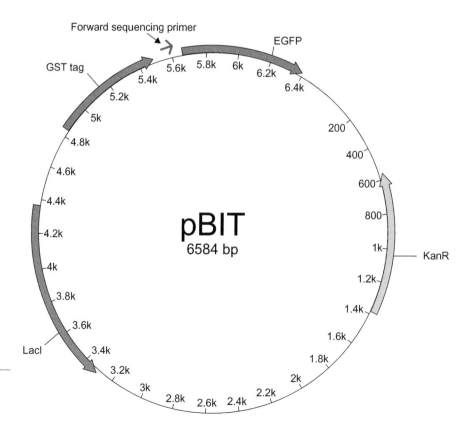

FIG. 9.3

Schematic of pBIT showing the location of the sequencing primer.

Next week, you will analyze the result from your sequencing reaction to determine the sequence of *egfp* in your clone.

Laboratory Exercise

Note: The instructions below are for preparing your sequencing sample for submission to the GENEWIZ sequencing facility. There are many different sequencing facilities available throughout the world, with many academic institutions having their own facilities. If you use a different facility, follow their instructions.

1. Obtain a thin-walled PCR tube in which you will set up your sequencing reaction. Your TA/instructor will give you details on how to label the tube.
2. Choose which clone you are going to sequence based on your previous results. Choose DNA from a colony that was positive using all of the different screening methods, if possible.
3. Using your miniprepped DNA (purified in Session 8C), determine the concentration of your DNA sample (as explained in Lab Session 2).
4. The sequencing reaction should contain 500 ng of your plasmid DNA and 25 pmol of primer. (**Note**: If your DNA has too low a concentration to obtain 500 ng, try choosing a different clone. If all of your clones have a low concentration, choose the clone with the highest concentration and use as much as possible.) Set up your reaction as follows:

DNA template (500 ng)	_____ µl
Forward primer (10 pmol/µl)	2.5 µl
Water	Bring to 15 µl total

   Remember, for a sequencing reaction, you only add ONE primer! (You aren't performing PCR and amplifying your template; you are making partial copies of one strand.)
5. Tap the tube against the bench to be sure reagents mix and are at the bottom of the tube.
6. Give your tube to your TA/instructor for submission to GENEWIZ or other sequencing facility.

Now that you have performed several different techniques for screening transformants, it is interesting to compare the reliability of each method. Use the following chart to do so. Leave blank the spaces for transformants that were not tested in specific assays. Keep in mind that the fluorescence test was the most reliable assay method, so use it as your reference for determining which assays gave the most consistent results.

Transformant number	mAB probe	PCR screen	Restriction digestion	Fluorescence assay

References

1   Sanger F, Nicklen S, Coulson AR. DNA sequencing with chain-terminating inhibitors. *PNAS*. 1977;74(12):5463–5467.
2   Butler JM. *Fundamentals of Forensic DNA Typing*. San Diego: Elsevier/Academic Press; 2009.
3   Dolan DNA Learning Center: Biology Animation Library, "Cycle Sequencing". Cold Spring Harbor Laboratory.

Discussion Questions

1. Will the restriction mapping screen be able to detect transformants with multiple inserts? Why?
2. If your DNA was not cut to completion during the restriction analysis of your transformants, what would you expect your gel to look like? Would you still be able to distinguish positive from negative clones?
3. When performing DNA sequencing, you only use one primer per reaction. What would happen if you mistakenly added both a forward and reverse primer to a sequencing reaction? How would your data be affected?

Computational Analysis of DNA Sequence from a Positive Clone: Part 2

Goal: You will analyze the data obtained from the sequencing reactions from last week to ensure your clone has the proper *egfp* sequence.

Introduction

Automated DNA sequencing is a frequently used tool to determine the sequence of DNA in a given sample. While DNA sequencing has evolved from its inception to be an automated process, the data obtained from the computer still requires human validation. Indeed, there are many times the computer analysis of the readout is ambiguous, leaving you to interpret the data and decide the actual sequence of the DNA.

In sequencing by chain termination (Sanger sequencing) using fluorescent dideoxynucleotide triphosphates (ddNTPs), the data is obtained in the form of a fluorescent chromatogram. Each peak represents a synthesized DNA fragment with the color representing which ddNTP was incorporated as the chain terminator. As such, the peaks on the chromatogram are read out sequentially (from smallest to largest, starting closest to your sequencing primer) to determine the sequence of your DNA molecule.

Chromatograms are typically saved as .abi files. There are many different programs which can view these files. The instructions in this lab session are specific for Chromas Lite, which is available as freeware via Technelysium Pty Ltd (http://www.technelysium.com.au/chromas_lite .html). Upon opening your chromatogram, you'll see a trace similar to that shown in Figure 10.1, which is a chromatogram for the pBIT sequence. The box shows the slider used to change the horizontal resolution of the trace, while the circle is used to change the vertical resolution.

In looking at your chromatogram, there are a few different things to consider. First, remember from Lab Session 9 that the interpretable sequence begins 15–30 nucleotides downstream of the end of the sequencing primer. As shown in Figure 10.2, the first ~25 peaks of the trace are rounded and not well resolved compared to the peaks that follow. So despite the fact that DNA fragments will be synthesized starting from the nucleotide

Molecular Biology Techniques. DOI: 10.1016/B978-0-12-385544-2.00010-7

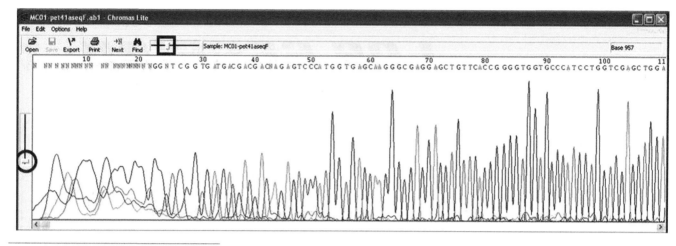

FIG. 10.1

Sample image of the first ~110 nucleotides of a chromatogram. The box indicates the bar to adjust the horizontal resolution and the circle represents the bar to adjust the vertical resolution.

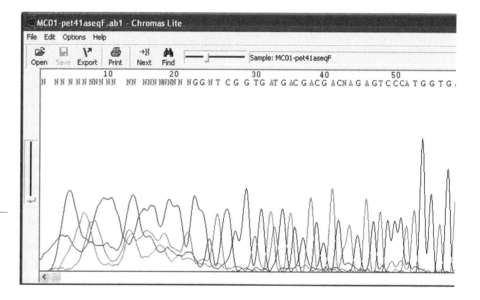

FIG. 10.2

pBIT sequencing data at the beginning of the chromatogram. Note the broad, rounded peaks in the beginning transitioning to sharp, evenly spaced peaks.

immediately after the end of the sequencing primer, you can see from this example why the sequencing primer needs to be designed upstream of the sequence you wish to sequence. Our sequencing primer is designed to anneal 60 nucleotides upstream of the junction between the *egfp* insert and pET-41a. You can see that by nucleotide 30, the peaks have become well defined and equally spaced, and there is very little noise at the bottom of the trace. This is where the sequence can begin to be read from.

Figure 10.3 shows a screenshot of what the chromatogram should look like towards the middle of the sequencing read. You can see that during this part of the trace the peaks are very well defined with very little noise present. The peak height in this region is often much higher than the peaks at the beginning of the trace. Most importantly, this sequence is very "clean;" in other words, it is unambiguous with regard to the sequence it represents. Looking above the peaks in the chromatogram, you will see that the software reads the chromatogram and assigns a nucleotide to the corresponding peaks. In the region shown, the clarity between the peaks allows the computer to do so very accurately. Looking back at Figure 10.2, you will

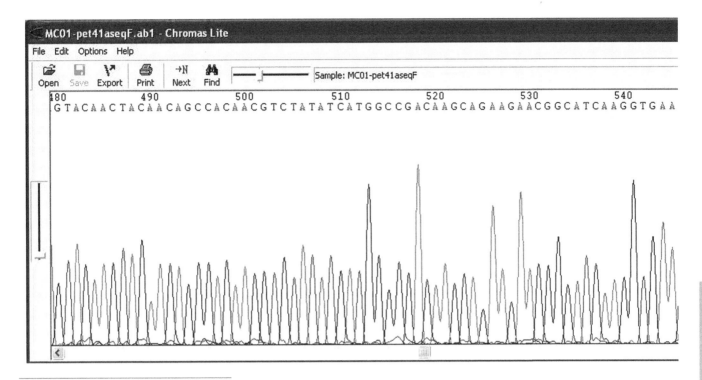

FIG. 10.3

pBIT sequencing data at the middle of the chromatogram. Note the sharp, evenly spaced peaks throughout and the lack of Ns in the software-generated sequence above the chromatogram data.

notice the beginning of the read contains a number of "N" bases, which means that the software was unable to confidently assign a base to a given peak. In the middle part of the sequence (Figure 10.3), Ns should be very infrequent, if present at all.

Each sequencing reaction will yield a different length, depending on the quality of the DNA and other factors. Generally speaking, a typical sequencing read will span 800–1000 nucleotides of unambiguous sequence (like that shown in Figure 10.3) before the signal starts to deteriorate. Eventually, the peaks become broad, more rounded and less evenly spaced, making assigning the proper base more difficult. As shown in Figure 10.4, the peak height has diminished and there are a few Ns that have crept into the sequencing read. This sequence can still be usable if you are able to confidently assign the bases by visually inspecting the chromatogram, but in general, if there are ambiguous bases such as these, sequencing with a different primer that would place these nucleotides in the middle of the sequencing run would be preferred.

In our case, we are sequencing the *egfp* gene that was cloned into pET-41a to be certain that no mutations were made to *egfp* during PCR cloning. While some of the sequence obtained is ambiguous, we will be able to answer our question as long as the *egfp* sequence is clean. Since we know that *egfp* was cloned into the *Nco*I site (CCATGG) at the 5′ end, we can scan the sequence and look for this site. Looking back at Figure 10.2, you can see at nucleotide 50 the sequence reads CCATGG, depicting the *Nco*I site and the beginning of the *egfp* insert sequence. Knowing that the *egfp* insert is about 720 nucleotides long, this means that the sequence should span from about nucleotide 50 to nucleotide 770. In Figure 10.4, we see that the signal doesn't deteriorate until approximately nucleotide (nt) 960. Thus, we should have an adequate read to look at our sequence.

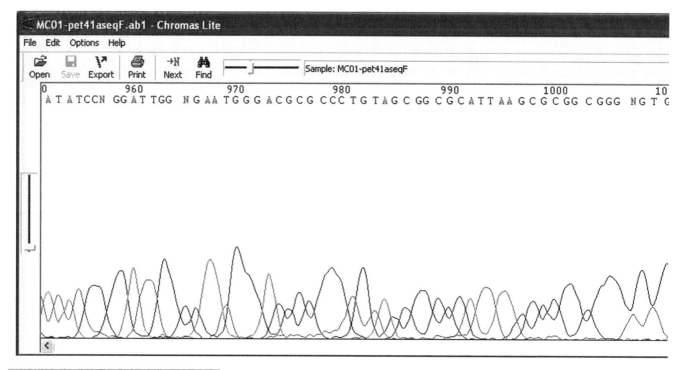

FIG. 10.4

pBIT sequencing data at the end of the chromatogram. Similar to the beginning of the chromatogram, broad, rounded peaks begin to show up leading to ambiguities, depicted by the Ns in the base called sequence.

Scanning through your sequence of interest (in our case, approximately nucleotides 50–770), you want to see if there are any Ns that were assigned by the software. You can scan through the sequence manually, as well as by using the "Next N" button that is circled in Figure 10.5, left panel. If there are Ns present, you want to use your best judgment to assign these bases. While there are no Ns found within the region of interest in the sample performed, Figure 10.5 (left panel) shows an example of an N found outside this range, with the horizontal and vertical resolution changed to better view the chromatogram at this exact point. What you can see at base 829 is the presence of a shoulder between the A at nt 828 and the G at nt 830. It appears as though this might be a doublet of consecutive Gs; however, you can compare this doublet to others in the chromatogram, such as those shown in the right panel (bases 708–709; 711–712). If you were determining the DNA sequence of a completely unknown sample, you would want to rely on data from a different primer or overlapping sequencing read to have confidence in assigning position 829. In our case, we rely on making an educated guess using the known *egfp* sequence to help us. If the known sequence of *egfp* in the region corresponding to nucleotides 826–831 in the chromatogram was CTAGGC, we could be relatively confident in assigning position 829 as a G. Similarly, if the sequence of *egfp* in that region was CTAGC, we could reasonably assume that despite the visual shoulder, the sequence only contained one G. One of the main differences between verifying a known sequence and determining an unknown DNA sequence is that during verification, one can make this kind of educated guess.

Ambiguous nucleotides in a chromatogram can appear for a variety of reasons. In addition to the typical decrease in sequence quality at the beginning and end of a sequencing reaction, other factors can lead to

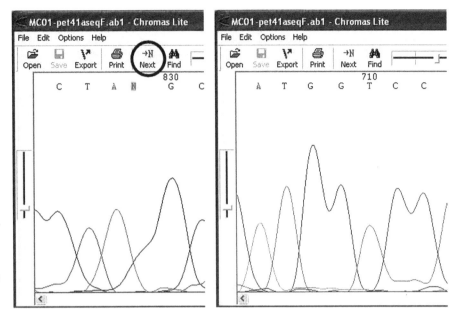

FIG. 10.5

Zoomed in view of the pBIT chromatogram. Note in the left panel an N at position 829 due to a shoulder in the trace. In the right panel, examples of traditional doublets (not shoulders) are shown at positions 708–709 and 711–712.

difficult-to-interpret chromatograms. If there are Ns throughout your entire sequence, this is likely due to sub-optimal conditions for your sequencing reaction. Low quality or insufficient quantity of DNA or poor quality sequencing primer (low T_m or mismatch with template) are the most common explanations for an overall poor-quality sequencing result. Even within good sequencing reads, ambiguities sometimes arise. One of the most common reasons for ambiguities is a stretch of a single consecutive nucleotide. These can often be problematic, both in deducing the correct number of bases in that stretch, and also in complicating the sequence immediately following the homopolymeric region (i.e., a poly-A region of a eukaryotic cDNA). There are some sequencing "tricks" a sequencing facility can perform to obtain higher quality sequencing reads if this type of sequence is expected in a given sample.

Comparing your sequencing data to the known *egfp* sequence can be performed a number of different ways. There are sequence alignment tools available on the Internet, such as ClustalW and many others. Another simple method to compare your sequence data to known sequences is by using the Basic Local Alignment Search Tool (BLAST) available from NCBI (www.ncbi.nlm.nih.gov/BLAST). BLAST is a frequently used tool for calculating similarity between sequences.[1] Applications of BLAST include studying evolutionary differences between genes or proteins in different organisms, predicting functions of new genes based on homology to genes of known function, and identifying particular motifs or domains within a sequence. Instructions for comparing your sequencing data to the known pBIT sequence using BLAST are described below.

Laboratory Exercise

1. Open your .abi file in Chromas Lite. Open the software, click the Open button, then find and open the file containing your sequence.

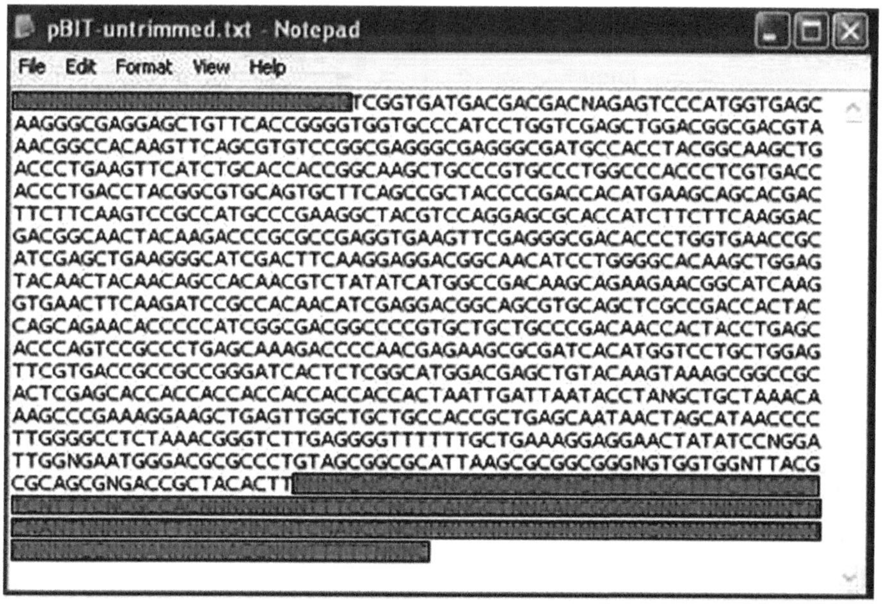

FIG. 10.6

Example of trimming sequence before BLAST search. Highlighted in red are regions of the sequence that are rich in Ns, meaning that the sequencing reaction gave ambiguous results in these regions. These regions are deleted before continuing to BLAST.

2. Copy your sequence to a text editor program and trim beginning and trailing Ns from the sequence. In Chromas Lite, click "Edit", "Copy Sequence" and choose "Plain Text". This will copy the entire sequencing read. Next, paste the sequence into Notepad, or a comparable text-editing program. Finally, in the beginning and end of the sequence, you will likely see many Ns, which are nucleotides that the sequencer was unable to accurately call (or identify) during the sequencing run. Simply delete the sequence containing multiple Ns in the beginning and end of the sequence. There will occasionally be a few random Ns in the middle of the sequence; you don't need to worry about these. An example is shown in Figure 10.6. In this case, the region highlighted in red would be deleted from your sequence before continuing to BLAST.

3. Perform a BLAST search on your trimmed sequence. To perform a BLAST search, go to http://blast.ncbi.nlm.nih.gov/Blast.cgi and select nucleotide BLAST. At your next window, copy and paste your modified/trimmed sequence into the box where it says "Enter accession number(s), gi(s), or FASTA sequence(s)." Under "Choose Search Set/Database," use the drop-down menu to select "Nucleotide Collection (nr/nt)." Leave the rest of the options at their default settings and click the "BLAST" button at the bottom of the screen. Figure 10.7 depicts the relevant regions of the BLAST webpage.

4. Analysis of BLAST sequence. After a few seconds, you should have your BLAST results. At the top of the page, there will be information relevant to your search (query length, database searched, etc.). A bit further down, a graphic summary will be presented and beneath that a tab saying "Descriptions" should appear. The top hit under Descriptions should be pBIT (Accession # JF275063.1). As shown in Figure 10.8, there are a few different headings. The E value (or Expected Value) is a measure of the significance of the match. The closer this number is to zero, the more significant the match. In order to view the actual sequence alignment between your input sequence and the known pBIT sequence, click on the max score, as shown in the circle in Figure 10.8.

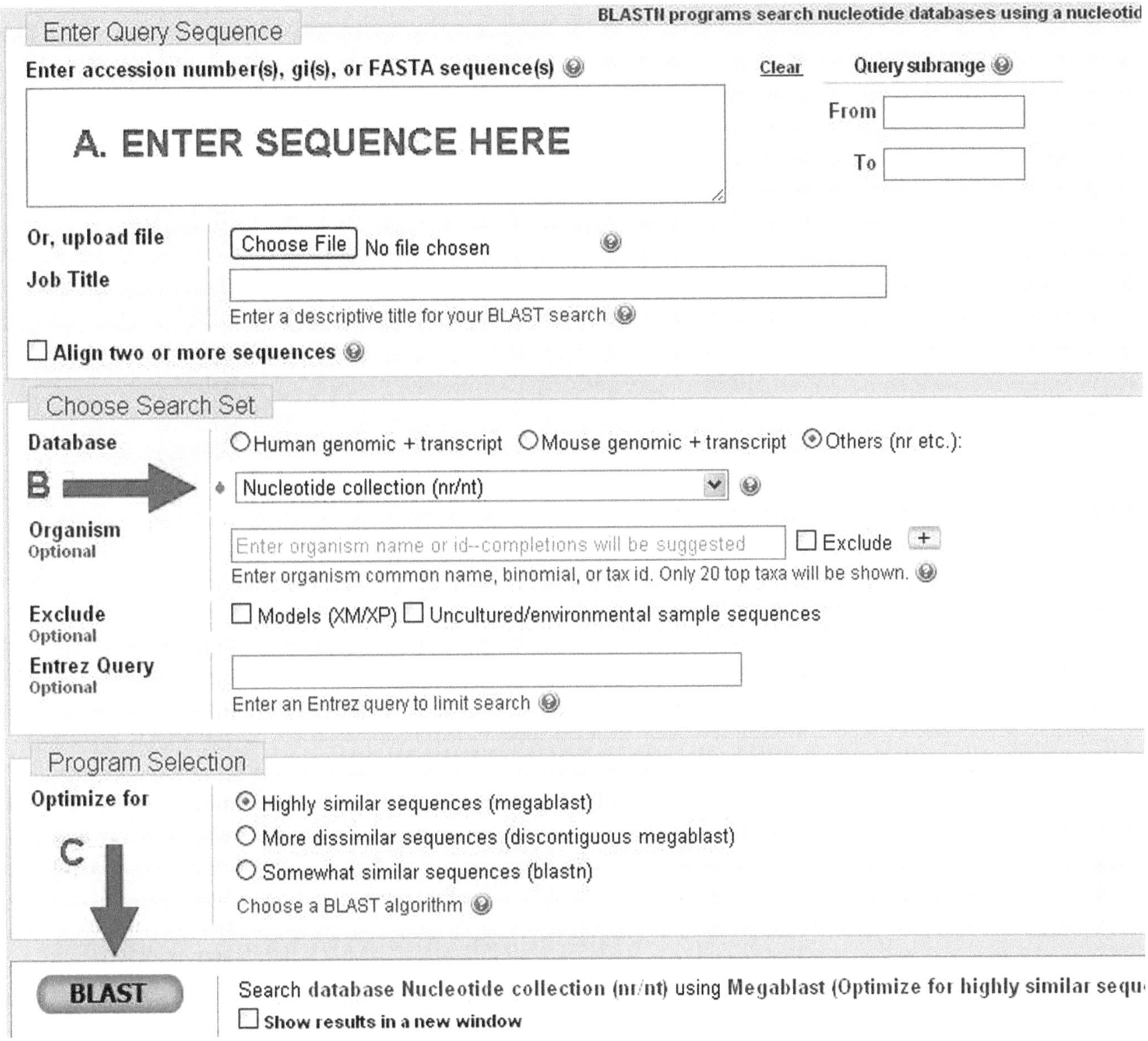

BLASTN programs search nucleotide databases using a nucleotid

Enter Query Sequence

Enter accession number(s), gi(s), or FASTA sequence(s) ⓘ   Clear   Query subrange ⓘ

A. ENTER SEQUENCE HERE   From [   ]

                                                                               To [   ]

Or, upload file   [Choose File] No file chosen   ⓘ

Job Title   [   ]
                   Enter a descriptive title for your BLAST search ⓘ

☐ Align two or more sequences ⓘ

Choose Search Set

Database   ○ Human genomic + transcript ○ Mouse genomic + transcript ⊙ Others (nr etc.):
B ➡   ◆ [Nucleotide collection (nr/nt)   ▾] ⓘ

Organism   [Enter organism name or id--completions will be suggested] ☐ Exclude [+]
Optional   Enter organism common name, binomial, or tax id. Only 20 top taxa will be shown. ⓘ

Exclude   ☐ Models (XM/XP) ☐ Uncultured/environmental sample sequences
Optional

Entrez Query   [   ]
Optional   Enter an Entrez query to limit search ⓘ

Program Selection

Optimize for   ⊙ Highly similar sequences (megablast)
C ⬇   ○ More dissimilar sequences (discontiguous megablast)
                ○ Somewhat similar sequences (blastn)
                Choose a BLAST algorithm ⓘ

[BLAST]   Search database Nucleotide collection (nr/nt) using Megablast (Optimize for highly similar sequ
                ☐ Show results in a new window

FIG. 10.7
BLAST search page. (A) Box where you enter your trimmed sequence. (B) Drop-down box to select Nucleotide collection (nr/nt) database. (C) BLAST button.

This should take you to the alignment of your sequence next to pBIT (Figure 10.9). You want to look specifically to make sure that the *egfp* insert you PCR amplified and cloned has the correct sequence. If you remember, the PCR product contained an *Nco*I site (CCATGG) on the 5′ end and a *Not*I site (GCGGCCGC) on the 3′ end. In pBIT, the *Nco*I site can be found starting at nucleotide 5674, and the *Not*I site finishes at nucleotide 6404. These regions are boxed in Figure 10.9.

In Figure 10.9, matches are shown by a line between the query (your input) and subject (pBIT). In this example, the match is 100% in the *egfp* region and shows that we indeed cloned a non-mutated form of our insert. If the sequence isn't a match, there would be no line between the two. You will notice that a couple Ns were still in our search (query positions 18, 804 and 932), which do not have a line with the corresponding nucleotide in the pBIT sequence. These positions are outside the

Sequences producing significant alignments:

Accession	Description	Max score	Total score	Query coverage	E value	Max ident
JF275063.1	Synthetic construct plasmid pBIT GST/EGFP fusion protein ger	(1712)	1843	99%	0.0	99%

FIG. 10.8
BLAST results.

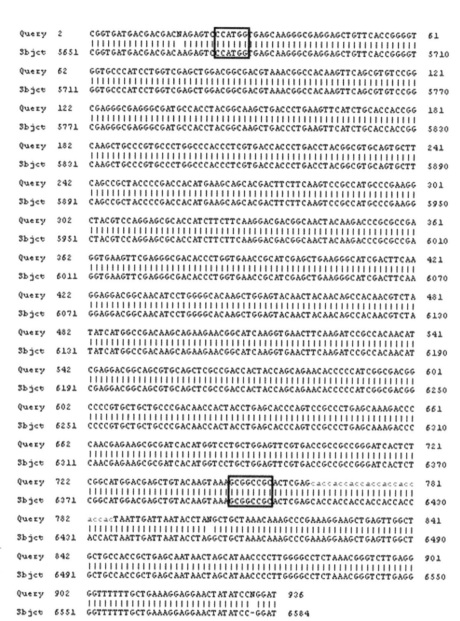

FIG. 10.9
Sequence alignment between the input sequence and pBIT. The *egfp* sequence is the region between the boxes depicting the *Nco*I (CCATGG) and *Not*I (GCGGCCGC) sites.

egfp reading frame, so they don't affect our ability to confirm the *egfp* sequence. However, if there were Ns in the *egfp* sequence, you would want to go back to the chromatogram and further examine the data to see if the sequence appears to match. If multiple Ns were present in the sequence, in practice you would probably sequence the region again (perhaps with a different sequencing primer) to obtain more data to accurately assign the ambiguous bases.

In most cases, your sequencing data will be a perfect match with pBIT. However, on the rare occasion that PCR introduces a mutation into the *egfp* insert sequence, this would enable us to detect the mutation and act accordingly. If the mutation was a silent mutation (a nucleotide change that does not alter the protein coding sequence), your experiment should not be affected. On the other hand, if the mutation incorporates a missense or a nonsense mutation, you would want to go back and choose a different positive clone to continue with your experiment.

Discussion Questions

1. Why is only the DNA corresponding to *egfp* sequenced in this lab exercise? Why do we not anticipate any mutations occurring outside the *egfp* sequence?
2. When doing your BLAST search, what do you notice about the other hits that show up as significantly similar to your sequence?

Reference

1   Madden T. The BLAST Sequence Analysis Tool. In: McEntyre J, Ostell J, eds. *The NCBI Handbook*. Bethesda, MD: National Center for Biotechnology Information; 2003.

PART 3

Expression, Detection and Purification of Recombinant Proteins from Bacteria

In this portion of the course, you will learn how to purify a genetically engineered protein from bacteria. As we have previously discussed, many pharmaceuticals and industrially-relevant enzymes are produced using recombinant methods. Furthermore, as molecular biologists, we are interested in the events that govern the life of a cell. How is the cell cycle controlled? What happens when a developing cell begins to differentiate? What are the factors that control cell growth? Why and how do cells die? If the organism is a pathogen, how is it able to infect the host? Before we can answer these questions in detail, we must know more about the actual molecules that participate in these processes. One classic biochemical solution to this problem is to take the cell apart and isolate each of its components for study. For proteins that are highly expressed in the cell, this approach is generally feasible. However, nothing short of herculean efforts have been necessary to purify protein or peptides that are present in extremely small quantities. This represents an important limitation, since highly active gene products such as enzymes or hormones are typically present at very low levels in biological materials.

With the advent of recombinant DNA techniques and the abundance of sequence information, it became much easier to clone the gene and express it in bacteria than it was to purify the protein from the native organism. You have now finished cloning the *egfp* gene into the pET-41a expression vector. As you will see, the GST::EGFP fusion protein may constitute as much as 10% of the total bacterial protein. The abundance of the fusion protein reduces the proportion of contaminating protein, but it is still necessary to have a method to fractionate it. Many of the pET vectors have been engineered to allow a one-step fractionation method called affinity chromatography, which can be used to isolate pure protein in relatively large quantities.

In this part of the course, you will induce the expression of the fusion protein from your plasmid clone. Following the induction of your bacterial cultures, you will visualize the protein expression pattern by SDS-PAGE and western blot. After confirming high levels of recombinant protein expression, you will grow a larger scale culture for purification by affinity chromatography. You will conduct fluorescence assays to monitor the quantity of the fusion protein in different stages of the purification procedure and to ensure the fusion protein is correctly folded and retains functional integrity. The purity of the final preparation will be analyzed by SDS-PAGE.

LAB SESSION 11

Expression of Fusion Protein from Positive Clones, SDS-PAGE and Western Blot: Part 1

Goal: This week, you (or your instructor) will induce small-scale cultures of your clones with isopropyl-β-D-thiogalactopyranoside (IPTG) and prepare protein samples for gel electrophoresis.

We will harvest actively growing, logarithmic-phase bacterial cultures to prevent proteolysis of the fusion protein, which can occur during the stationary phase of growth. IPTG is added just a few hours before the bacteria are harvested since the induction of recombinant protein is often detrimental to bacterial growth and viability.

Today, you will make lysates of the cultures, separate them by Sodium Dodecyl Sulphate-Polyacrylamide Gel Electrophoresis (SDS-PAGE), and transfer to a nitrocellulose membrane. You will save the nitrocellulose membrane for next week to finish the western blot.

Molecular Biology Techniques. DOI: 10.1016/B978-0-12-385544-2.00011-9

LAB SESSION 11A

Interim Laboratory Session

Laboratory Exercise

Inoculate Cultures for SDS-PAGE

1. Heavily inoculate three 1 ml LB/kan cultures with positive clones from the master plate that you incubated most recently. Use a separate toothpick to inoculate each of the LB/kan tubes.
2. Label the three LB/kan cultures with your lab day, station number and the transformant number.
3. Place the inoculated cultures in designated racks.
4. The LB/kan tubes containing your inoculated cultures will be refrigerated to retard growth. The afternoon before your lab period, they will be placed in a shaker at 37°C and incubated overnight along with positive and negative controls prepared by the instructor.

LAB SESSION 11B

Expression of Fusion Protein from Positive Clones, SDS-PAGE and Western Blot

Introduction

A western blot is similar to a colony hybridization using a monoclonal antibody probe, except that isolated proteins are first separated by molecular weight and then transferred to nitrocellulose.

SDS-PAGE is used to separate proteins by molecular weight. SDS-PAGE of proteins has numerous applications, including molecular weight determination, determining sample purity, quantifying expression, immunoblotting (western blot), and isolating proteins for peptide sequencing or for generating antibodies.

You are already familiar with DNA agarose gel electrophoresis, and SDS-PAGE does share some similarities with this method. Both methods separate molecules by size, use electrical charge differences to cause migration and both require a matrix to separate molecules by size. There are several differences between the two types of electrophoresis.

1. DNA is routinely separated on agarose gels, while proteins are generally run on polyacrylamide gels. This is because proteins are generally smaller than DNA and polyacrylamide matrices have a smaller pore (sieve) size than agarose.
2. By convention, DNA is run at a constant voltage and protein is run at a constant current.
3. All DNA is negatively charged, but proteins have varying charges depending on the amino acid content of the specific polypeptide and the pH of the buffer. Some proteins are positively charged, while some carry a net negative charge.
4. DNA, especially linear DNA, has little secondary structure, while proteins can be globular or linear, and can form dimers and other multimers.

Because of #3 and #4 above, if proteins were run on a native or non-denaturing polyacrylamide gel (i.e. run without SDS), protein migration would depend on at least three factors: size, charge and shape.

SDS-PAGE allows proteins to migrate by size alone, through the use of SDS and a reducing agent. SDS is an ionic detergent that denatures (unfolds) proteins by wrapping around the polypeptide backbone forming a micelle, and thus conferring a net negative charge in proportion to polypeptide length. SDS also disrupts most non-covalent bonds, thereby decreasing protein folding.

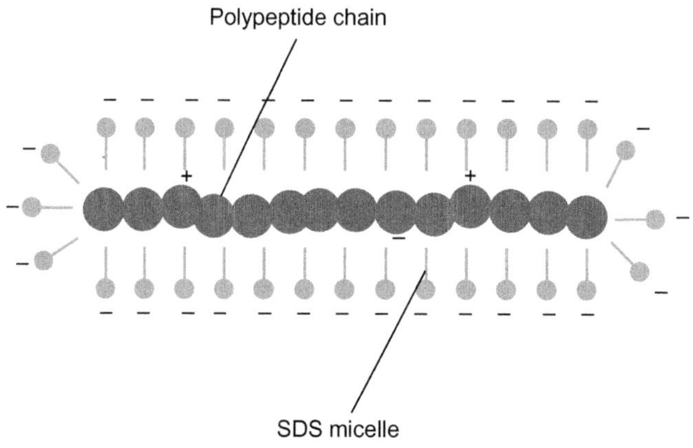

Polypeptide chain

SDS micelle

FIG. 11.1

Polypeptide treated with β-mercaptoethanol and SDS. β-mercaptoethanol breaks the covalent disulfide (cysteine) bonds. SDS disrupts non-covalent interactions, further denaturing the protein, and forms a micelle around it, conferring a negative charge.

A reducing agent such as β-mercaptoethanol or dithiothreitol is added to reduce cystine bonds (disulfide bonds) and further unfold the proteins.

After boiling a protein sample in SDS and β-mercaptoethanol, all proteins act as negatively charged linear molecules and can be electrophoretically separated by size alone (Figure 11.1).

After running the gel, it can either be stained non-specifically to visualize the protein bands using Coomassie Blue, GelCode Blue, or silver stain; or the proteins can be transferred to a nitrocellulose membrane for immunoblotting (western blotting) to visualize a specific protein of interest. In Lab Session 14 we will run an SDS-PAGE gel and stain it using GelCode Blue to visualize protein bands.

In today's lab session, we will perform a western blot (to be completed in the following laboratory session). The first step of this process is to transfer the proteins from the polyacrylamide gel to the nitrocellulose membrane. Total protein on the nitrocellulose membrane may be visualized at this point using the water-soluble Ponceau stain. After the proteins are transferred, probing the blot is very similar to the protocol performed in Lab Session 7A "Colony Hybridization with an EGFP Monoclonal Antibody probe," except the protein on the membrane was derived from the polyacrylamide gel, rather than the colony lift. This portion of the western blot will be completed in the next laboratory session.

The molecular weight of the GST::EGFP fusion protein can be estimated, assuming the average weight per amino acid is equal to 114 Da. The *gst* gene is 660 bp, encoding 220 amino acids: $220 \times 114 = 25,080$ Da. The *egfp* gene is 720 bp, encoding 240 amino acids: $240 \times 114 = 27,360$ Da. There are 174 additional nucleotides between the *gst* and *egfp*, encoding 58 amino acids: $58 \times 114 = 6,612$ Da. The weight of the fusion protein can therefore be approximated as: $25,080$ Da $+ 27,360$ Da $+ 6,612$ Da $= 59,052$ Da or ~59 kD. You can determine the actual molecular weight using free online software; the actual molecular weight of our fusion protein is 58,500 Da. You will be able to visualize this band in your positive clones using the Ponceau stain and then confirm that it is indeed the correct band at the completion of the immunoblot exercise next week. In the negative clones, you may see a band of approximately 25 kD, corresponding to the GST protein alone (with nothing fused to it).

Laboratory Exercise

SDS-PAGE and Western Blot: Part 1

GEL APPARATUS PREPARATION

Note: The instructions for assembling the gel apparatus and transblot assembly are specific for the Bio-Rad Mini-Protean 3 Electrophoresis unit and the Bio-Rad Mini Trans-Blot Cell. If you are using equipment made by another manufacturer, your instructor will provide you with assembly and usage instructions.

We are using precast polyacrylamide gels, so you don't have to pour them. If you must pour your own polyacrylamide gels in the future, keep in mind that non-polymerized acrylamide in both the powder and liquid forms are potent neurotoxins. Great care must be taken to avoid inhalation, ingestion and skin-exposure to this substance. Solidified polyacrylamide does not pose the same safety hazard.

1. Two gels must be run per unit, so you need to pair with another group.
2. Assemble the Bio-Rad apparatus as demonstrated by your instructor, with the short plates toward the center. If you are not paired with another group/student and only have one gel to run in the apparatus, you will need to use a buffer dam in place of a second gel.
3. Add Tris-glycine running buffer to the top of the center buffer chamber (be sure the buffer covers the wells) and also to the outer chamber (only about 2 inches deep). Check for leaks.
4. Wash wells with Tris-glycine running buffer.

SAMPLE PREPARATION

Steps 1–3 will be done for you.

1. Add 2 ml of 2× YT/kan/IPTG medium to each overnight culture.
2. Also inoculate positive (pBIT) and negative (pET-41a) controls into 2× YT/kan/IPTG medium for each lab group.
3. Incubate at 37°C for 3–4 hours to induce protein expression.
4. Set heat block to 95°C or higher (be sure heat block has water in the holes for heat transfer).
5. Harvest cultures. Spin 0.75 ml of each culture for 2 minutes in a microcentrifuge (use locking or screw-top microfuge tubes, if available). Also, prepare two additional samples: pBIT and pET-41a which have been grown in 2× YT/kan/IPTG medium for you. Remove and discard supernatants, saving pellets in tube. Invert tubes over paper towels to remove residual liquids.
6. To each tube, add 100 µl dH$_2$O and vortex until pellet is dispersed.
7. To each tube, add 100 µl of 2× sample buffer. DO NOT VORTEX TO MIX. 2× sample buffer has SDS which will foam when vortexed. Mix by gently pipetting up and down.
8. Cap the tubes firmly and use tube locks to secure caps shut. Incubate at a minimum temperature of 95°C for 10 minutes (be sure water is in heat block holes – incubating in boiling water is fine).
9. Microcentrifuge the tubes at maximum speed for 15 minutes.

93

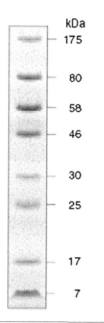

FIG. 11.2

The New England Biolabs prestained broad range molecular weight marker in kilodaltons.

10. Carefully transfer about 50 μl of supernatant from each tube to a new labeled tube. Be careful not to transfer any pellet material. Although the samples can be stored at −20°C, we will perform SDS-PAGE today. Store them on ice until the gel is ready.

PREPARING MOLECULAR WEIGHT STANDARDS

We will be using the New England Biolabs prestained broad range molecular weight marker (Figure 11.2). The molecular weight marker should already be mixed with loading buffer. When you are ready to load the gel, make sure the tube of molecular weight standard is securely capped and heat it for about 1 minute at 95°C. Place on ice until you are ready to load it.

LOADING SAMPLES ON THE GEL

1. To practice, use a micropipette tip to add 2.5 μl of 2× sample buffer (containing bromophenol blue) to a lane. Do not try to expel all of the liquid in the tip: an air bubble will result, which could distort the migration of your sample and could also force your sample out of the well. Practice until you feel comfortable loading the 2× sample buffer.
2. Add 8 μl of molecular weight marker (MWM), 10 μl of the positive control (pBIT), 10 μl of the negative control (pET-41a) and 10 μl of each transformant fusion protein sample (from tubes 1–3) to individual lanes according to the following scheme. Do not load sample into the first or last lane (you can practice in them).

2	3	4	5	6	7
MWM	pBIT	pET-41a	Sample 1	Sample 2	Sample 3

ELECTROPHORESIS

1. Two gels will be run per apparatus: coordinate with another group before you start.
2. Attach the electrodes to the power supply.
   CAUTION: Be sure to match the anode (red) with the positive lead and the cathode (black) with the negative lead.
3. Activate your power supply. Run your gel at 30 mA constant current. If you are running two gels on a single unit (most of you will) the total current should be 60 mA. If there are two units per power supply, increase the current according to the number of additional gels (for example, 90 mA for three gels and 120 mA for four gels).
4. Run the gel until the bromophenol blue tracking dye reaches the bottom of the gel.

STOPPING ELECTROPHORESIS

1. Turn off the power supply.
2. Detach the leads from the power supply. Remove your gel from the apparatus. If the other gels connected to the power supply have not finished, turn the power supply back on, adjusting the current to the proper amperage for the number of remaining gels.

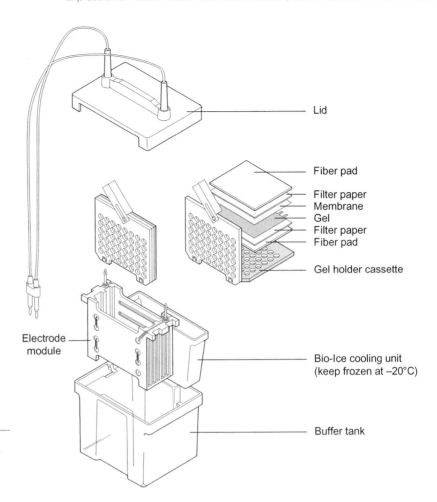

Lid

Fiber pad
Filter paper
Membrane
Gel
Filter paper
Fiber pad

Gel holder cassette

Electrode module

Bio-Ice cooling unit (keep frozen at −20°C)

Buffer tank

FIG. 11.3
Mini trans-blot cell description and assembly of parts. Copyright 2004 Bio-Rad Corporation. Used with permission.

3. Pour the buffer into the sink.
4. Disassemble the gel from plates as demonstrated.
5. Cut the wells off the gel after disassembling.

It is possible to stain the gel to visualize the proteins at this point, but for a western blot the gel cannot be irreversibly stained. Instead, we will first transfer the protein to a nitrocellulose membrane, and then stain the membrane with a water-soluble dye prior to completing the western blot.

TRANSFERRING TO NITROCELLULOSE

We will use the Bio-Rad Mini Trans-Blot Cell for transfer of protein from the polyacrylamide gel to the nitrocellulose membrane (Figure 11.3).

1. Prepare 1 liter of 1× Transfer Buffer by mixing 2.5 g of sodium bicarbonate into 1 liter of deionized water.[1]
2. Obtain a glass dish, two sheets of Whatman filter paper and one sheet of nitrocellulose. **Be sure to handle everything only when wearing gloves**.
3. Label nitrocellulose on top left in PENCIL.
4. Pour a small amount (about 1 cm deep) of 1× Transfer Buffer into the dish.
5. Thoroughly wet the nitrocellulose membrane, fiber pads and filter paper in the dish of transfer buffer. **Do not allow the nitrocellulose to touch the gel**.

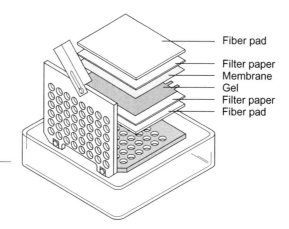

Fiber pad
Filter paper
Membrane
Gel
Filter paper
Fiber pad

FIG. 11.4

Close-up of "gel sandwich." Copyright 2004 Bio-Rad Corporation. Used with permission.

6. Prepare a gel sandwich as described below and depicted in Figure 11.4. Under no circumstances should you shift the position of the nitrocellulose on the gel.
   a. Place the cassette, gray side down, in the glass dish containing transfer buffer.
   b. Place one pre-wetted fiber pad on the gray side of the cassette.
   c. Place a sheet of filter paper on the pad.
   d. Place the gel on the filter paper. Remove bubbles by gently rolling a glass tube over it.
   e. Place the pre-wetted nitrocellulose membrane on the gel, labeled side down at the top of the gel. Remove bubbles by gently rolling a glass tube over it.
   f. Place a filter paper on the membrane. Remove bubbles by gently rolling a glass tube over it.
   g. Add the last fiber pad and close the sandwich.
7. Place the cassette in the module. The gray side of the cassette should face the black side of the module, and the clear side of the cassette should face toward the red side of the module.
8. Fill with transfer buffer to the top (if you run out, use the transfer buffer you used for soaking your gel).
9. Put on the lid of the assembly so that red matches red and black matches black.
10. Turn on the current to 300 milliamps for 40 minutes.

DISASSEMBLE AND STORE

1. Disassemble the apparatus.
2. Discard the gel and two sheets of filter paper, and rinse the fiber pads (fiber pads are NOT disposable) and the rest of the apparatus unit with plenty of tap water and put away.
3. Place the blot between two sheets of fresh Whatman filter paper and hand in to the instructor to be wrapped in foil and refrigerated at 4°C until next week.

Reference

1. Dunn SD. Effects of the modification of transfer buffer composition and the renaturation of proteins in gels on the recognition of proteins on western blots by monoclonal antibodies. *Anal. Biochem.* 1986;157:144–153.

Discussion Questions

1. Following SDS-PAGE, what is an advantage of protein detection by immunoblotting (western blotting) over a non-specific gel-staining procedure? Under what circumstances would it be desirable to non-specifically stain the gel, rather than immunoblotting?

2. The western blot is similar in procedure to the colony hybridization with an anti-GFP monoclonal antibody probe experiment we performed in Lab Sessions 6 and 7. Based on this and what you learned about western blot, how would the visualization of a blot differ from that of the colony hybridization experiment we did earlier? Does it give you more or less information about the protein being expressed? Is there a difference in the primary and secondary antibodies used in the two experiments?

LAB SESSION 12

Expression of Fusion Protein from Positive Clones, SDS-PAGE and Western Blot: Part 2

Goal: Today you will Ponceau stain your membrane to ensure an even transfer and complete the western blot that was started in the last laboratory session.

Introduction

Probing the western blot is analogous to the protocol of colony hybridization with a monoclonal antibody probe (Lab Session 7A), except that the protein on the membrane for the western blot was derived from the polyacrylamide gel, rather than the colony lift. Since the cellular proteins of each clone were separated by SDS-PAGE, you will be visualizing the specific protein band corresponding to the GST::EGFP fusion protein, rather than just seeing that the protein was made by a specific clone as in the colony lift experiment. For a review of how the GST::EGFP protein is detected on the membrane, revisit Figure 6.1. In brief, following protein transfer to the nitrocellulose membrane, the membrane is blocked with a protein that will not interact with the primary antibody (such as casein found in nonfat milk or bovine serum albumin). Primary antibody is added and allowed to bind to the specific epitope on the EGFP protein and unbound antibody is washed away. Secondary antibody conjugated to horseradish peroxidase (**g**oat **a**nti-**m**ouse **p**eroxidase; GAMP) is added and allowed to bind to the primary antibody, and then excess is washed away. Finally, the colorimetric substrate chloronaphthol is added. Protein samples that contain the GST::EGFP fusion protein will reveal a purple band with an apparent molecular weight of approximately 59 kD. Clones that did not express EGFP (negative clones) will show no band. Figure 12.1 shows a flowchart of the steps in performing a western blot.

Laboratory Exercises

SDS-PAGE and Western Blot: Part 2

PONCEAU STAIN

In addition to the western blot you will be performing, you will visualize the total protein band pattern for each sample. You are staining with a

Molecular Biology Techniques. DOI: 10.1016/B978-0-12-385544-2.00012-0

Western blot

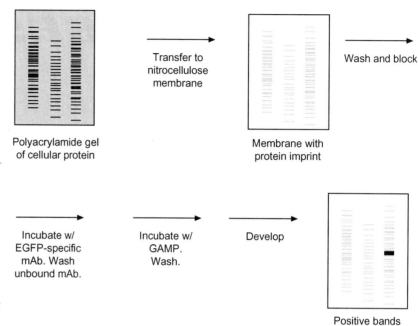

FIGURE 12.1

Western blot experimental flowchart. Samples are separated by SDS-PAGE, then transferred to a nitrocellulose membrane. The membrane is washed and blocked, then incubated with the primary antibody, α-EGFP. Excess primary antibody is washed off, and the secondary antibody bound to horseradish peroxidase (goat anti-mouse conjugated to horseradish peroxidase or GAMP) is added. Excess secondary antibody is washed off, then the blot is developed using the peroxidase substrate, chloronaphthol.

soluble dye called Ponceau Red that will be removed during the blocking step.

1. Remove the blot from the refrigerator.
2. Place the blot (labeled side up) in a square Petri dish and just cover (do not fill the dish) with Ponceau Red stain. Let it sit for 2 minutes.
3. After the 2 minutes (you will not see bands yet), rinse the blot briefly with distilled water. At this point you will see protein bands. You should see a band of approximately 59 kD molecular weight in the positive clones that is not present in the negative control or negative clones.
4. Record observations of the Ponceau stain in your notebook. *Were lanes evenly loaded; did any air bubbles appear on the blot?*

BLOCKING

The blocking step of the western blot is critical to prevent antibody from sticking non-specifically to the nitrocellulose membrane.

1. Pour the water out of the Petri dish and add 10 ml of blocking solution (TBS-T + 5% powdered milk).
2. Incubate for 5 minutes on an orbital shaker at 50 rpm.
3. Wash briefly in two changes of TBS-T.

INCUBATION OF THE BLOT WITH α-EGFP (PRIMARY ANTIBODY)

1. Obtain an aliquot of the α-EGFP (anti-EGFP). Store the antibody on ice.
2. Make a 1:1000 dilution of the α-EGFP by mixing 7.5 μl antibody with 7.5 ml TBS-T. Mix and add to the washed blot. Incubate on an orbital shaker for 1 hour.

3. Rinse twice with TBS-T.
4. Wash twice with fresh TBS-T (5 minutes each).

INCUBATION WITH GOAT ANTI-MOUSE PEROXIDASE (SECONDARY ANTIBODY)

1. Discard the final wash.
2. Make a 1:500 dilution of the goat anti-mouse antibody conjugated to horseradish peroxidase (GAMP) by mixing 15 μl with 7.5 ml of TBS-T. Add to the blot and incubate for 1 hour on an orbital shaker.
3. Rinse twice with TBS-T.
4. Wash twice with fresh TBS-T (5 minutes each).

COLORIMETRIC DETECTION OF PEROXIDASE ACTIVITY

1. Add about 10 ml of peroxidase substrate to the tray containing the blot. Color development (dark purple) should be evident within seconds to minutes.
2. Stop the reaction by rinsing with distilled water.
3. To preserve color, dry the blot between two pieces of filter paper.
4. Record the results in your notebook.

Positive clones should show a band with a molecular weight of approximately 59 kD, while negative clones should show no bands. In the positive clones, you may see some smearing or distinct bands below the 59 kD band. This is most likely due to partial degradation of the fusion protein.

Replica Plate Positive Clone

Replica plate one of your positive clones onto LB/kan so you will have a fresh inoculum for next week. Label, tape shut and place inverted in the 37°C incubator. Your instructor will save it in the refrigerator after an overnight incubation.

Discussion Questions

1. After performing the Ponceau stain, you may see a distinct band of approximately 25 kD in your negative control that is not present in your positive control or your other positive clones. What does this band likely represent?
2. If the proteins used in SDS-PAGE are denatured, how would the primary antibody bind to them? Is secondary structure not important for the antibody to bind, or are antibodies created to bind to the denatured version of proteins? *Outside resources will be necessary to think about this question.*

LAB SESSION 13

Extraction of Recombinant Protein from *Escherichia coli* Using a Glutathione Affinity Column

103

This week you will purify the GST::EGFP fusion protein by affinity chromatography. You will need to inoculate one of your clones into liquid culture at least one day before your lab. You or your instructor will induce protein expression with IPTG 3–4 hours before your regular lab period.

Molecular Biology Techniques. DOI: 10.1016/B978-0-12-385544-2.00013-2
© 2012 Elsevier Inc. All rights reserved.

LAB SESSION 13A

Interim Laboratory Session

Laboratory Exercise

Inoculate Cultures for Protein Purification

Inoculate one positive clone (heavily) into 2 ml LB/kan broth in a snap-cap tube. Do this by picking bacteria from your master plate with a sterile toothpick and dropping it into the tube containing LB/kan. Be sure to label the tube with your station number and lab day. If you are not completely confident that you have a positive clone, you should use pBIT as your clone for the protein purification. Place the tube in a designated rack.

Inoculated cultures will remain refrigerated until the evening before your lab, when your instructor will place them in the 37°C shaking incubator overnight.

LAB SESSION 13B

Extraction of Recombinant Protein from *Escherichia coli* and Purification Using a Glutathione Affinity Column

Goal: Today you will lyse bacterial cells that have been induced with IPTG to express the GST::EGFP fusion protein and prepare a crude cellular homogenate. You will then use affinity chromatography to purify the GST::EGFP fusion protein encoded by your clone. This is achieved by taking advantage of the affinity of the glutathione-S-transferase (GST) moiety of the fusion protein for the small molecular weight molecule, glutathione.

Introduction

The first step in purifying a cellular protein is lysing and homogenizing the cells. This will be accomplished by treatment with the enzyme lysozyme, followed by multiple freeze–thaw cycles and finally by sonication. Freezing and thawing bacterial cells lyses them by disrupting the cell membrane, releasing soluble periplasmic and cytoplasmic components. The mechanical action of sonication further disrupts the cells and separates cellular proteins and lipids. It also serves to shear the DNA. Following this homogenization of cellular components, our target protein, the GST::EGFP fusion protein, can be purified by the technique of affinity chromatography. The overall procedure for preparing cell lysate from your bacterial culture is outlined in Figure 13.1.

Affinity chromatography is a commonly used method for purifying recombinant proteins. The use of affinity chromatography has become widespread due to the development of numerous expression vectors containing sequences that may be used as affinity tags. These expression vectors contain DNA sequence encoding an affinity tag (also called a fusion tag) either directly upstream or downstream of the multiple cloning site. The gene of interest must be inserted in the correct orientation and reading frame with respect to the affinity tag. Once protein expression is induced, a fusion protein will be produced. The fusion protein will contain two polypeptide moieties; one corresponding to the affinity tag and one corresponding to the protein of interest. Numerous affinity tags exist. Some of the most common include glutathione-S-transferase (the tag we use), hexahistidine (a series of six histidine residues) and a cellulose binding domain.

pET-41a contains the gene for glutathione-S-transferase (*gst*) directly upstream of the polylinker region. Once a gene of interest (in our case, *egfp*)

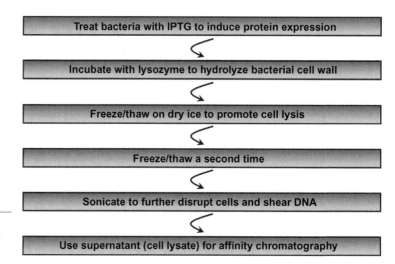

FIG. 13.1
Overview of preparing cell lysate from bacterial cells.

is inserted into the polylinker in the correct orientation and reading frame, a fusion protein can be expressed. The fusion protein we will express is GST::EGFP. We will take advantage of the GST moiety of the GST::EGFP fusion protein in order to purify it via affinity chromatography.

The principle of affinity chromatography relies on the binding of a biospecific ligand to the molecule of interest. The ligand for GST is glutathione, which is covalently attached to a support matrix to allow separation of glutathione binding molecules from other cellular components.

The principle steps of affinity chromatography are described below, and are illustrated in Figure 13.2. The specific details of our purification are indicated in italics.

1. **Injection of sample**. The sample contains a mixture of proteins and other cellular components along with the recombinant protein of interest.

   The sample you will use is a crude homogenate of the induced bacterial culture containing your GST::EGFP fusion protein.

2. **Adsorption of molecules with affinity for the ligand**. Molecules with specific affinity for the immobilized ligand will bind to the affinity resin while the rest of the molecules will flow through.

   Only the GST::EGFP fusion protein will bind to the glutathione-linked affinity resin.

3. **Wash impurities from the column**. Molecules with no (or weak, non-specific) affinity for the ligand are washed from the column.

   Cellular molecules other than the GST::EGFP fusion protein will be washed from the column, and the GST::EGFP fusion protein will remain bound to glutathione in the column.

4. **Elution of the target molecule(s) from the column**. A biospecific reagent binds to the affinity resin, dislodging and eluting molecules bound to the ligand.

   The GST::EGFP fusion protein will be eluted from the column using a solution of reduced glutathione.

The affinity column will be packed with a slurry we will refer to as the "affinity resin," illustrated in Figure 13.3. The substance that makes up

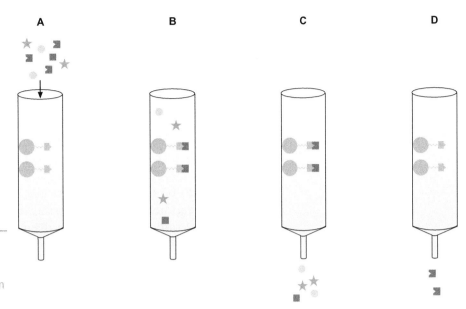

FIG. 13.2

The principle steps of affinity chromatography. (A) Sample injection. (B) Adsorption of target molecule(s). (C) Washing of impurities. (D) Elution of target molecule(s).

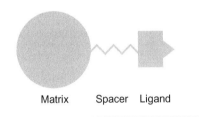

Matrix   Spacer   Ligand

FIG. 13.3

The affinity resin is composed of a biospecific ligand, a support matrix and a biochemically inert spacer.

the affinity resin has three covalently-bound components: the matrix, the spacer and the ligand.

The support matrix functions to suspend the ligand in the column. The matrix must be rigid, biochemically inert and have a high surface-to-volume ratio. The most commonly used support matrix is sepharose. Sepharose is composed of small agarose beads and is available in various sizes.

The spacer is a carbon chain that links the ligand to the matrix. The purpose of the spacer is to present the ligand to the molecule of interest. If no spacer was present, or if a spacer was too short, the ligand would be embedded in the matrix and might not be available for binding to the molecule of interest. Conversely, a spacer that is too long may have affinity for undesired molecules via hydrophobic interactions. The ideal spacer would be just long enough to present the ligand, but short enough to remain biochemically inert.

The biospecific ligand of the affinity resin specifically binds the molecule of interest. Depending on the type of molecule to be purified, various classes of ligands may be used. For example, if an enzyme is to be purified, the ligand may be a substrate, inhibitor, or cofactor of the enzyme. If an antibody is to be purified, the cognate antigen may be used as the ligand. If a nucleic acid is to be purified, ligands could be complementary nucleic acid sequences. (For example, an oligo-dT column is used to purify eukaryotic mRNA.) Affinity resins can be made in the laboratory by chemically crosslinking reactive resins with ligand, but you will use a commercially available affinity matrix. Because your protein is tagged with the enzyme glutathione-S-transferase, the biospecific ligand you will use is the enzyme's substrate, glutathione.

A final consideration when designing an affinity purification protocol is the issue of column capacity. Column capacity refers to the number of ligand molecules available for binding. In practice, this is related to the amount of resin used for purification. Different types of ligands will have different binding capabilities. If there is not enough ligand for all of the target molecules in the sample to bind, some of the target molecules will flow through the column and be lost. Conversely, if there is an excessive number of ligand binding sites compared to target molecules in the sample, non-target sample molecules with a lower affinity for the ligand may be able to bind to the affinity resin along with the target molecules. If these non-target molecules remain bound during the wash steps, they will be eluted together with the target molecules during the elution. This, of course, leads to impurities in the purified product. Optimally, the number of ligand binding sites would be exactly equal to the number of target molecules in the sample. In practice, this is impossible to achieve. Column capacity needs to be optimized empirically in most cases because expression levels of various proteins expressed from various promoters under varying growth conditions differ. The diagnostic SDS-PAGE you will be performing next week will help you analyze the purity of your sample. If the eluate does not appear to be pure, or if you lost a great deal of target protein in the washes, then you would modify your purification protocol either by adjusting column capacity or by adjusting the stringency of the washes.

Laboratory Exercises

Growing Bacterial Suspension Cultures for Fusion Protein Purification

Your instructor will inoculate your overnight culture into 100 ml of $2\times$ YT/kan/IPTG medium and incubate at 37°C for 3–4 hours.

Harvesting IPTG-Induced Cultures

1. Divide your culture (35–45 ml in each tube) between two large centrifuge tubes capable of withstanding $10,000 \times g$ force. Excess culture can be discarded in the "bacterial graveyard."
2. Pellet the bacteria by spinning for 5 minutes at $10,000 \times g$ in a high-speed centrifuge at 4°C.
3. Pour off the clear supernatant and discard.
4. Transfer the two pellets to a single microcentrifuge tube using the following method. Add about 0.75 ml of $2\times$ YT medium (or LB) to one bacterial pellet. Pipette and expel to re-suspend the pellet. Transfer this to the second pellet and pipette and expel to re-suspend. Now transfer the re-suspension to a 1.5 ml microcentrifuge tube. Repeat this procedure with 0.5 ml of $2\times$ YT to transfer any remaining bacteria.
5. Balance the weight of the bacteria-containing tube with that of another (water-containing) tube and spin for 1 minute in a microcentrifuge at 13,000 rpm to pellet the bacteria. Pipette off the clear supernatant. Invert the bacteria-containing tube over a paper towel and leave for about 5 minutes to remove residual liquid.

Breaking Open Bacterial Cells

1. Add 0.75 ml of ice-cold $1\times$ GST Bind/Wash Buffer (with Pefabloc, a protease inhibitor) to the pellet. Vortex to suspend the pellet.
2. Add 15 μl of lysozyme (10 mg/ml). Vortex and incubate on "wet" ice (H_2O) for 15 minutes.
3. Place the microcentrifuge tube on a bed of "dry" ice (CO_2) to freeze.

4. Thaw the tube by warming it in the palm of your hand. As soon as the sample melts, freeze it again on dry ice.
5. Thaw the tube as described above in step 4 and store it in liquid form on wet ice. Freezing and thawing helps release the fusion protein by disrupting the bacterial cell wall.

CAUTION: Put on protective ear guards before activating the sonicator in the next step.

6. You will use a sonicator with an immersible tip (probe). **Note**: There may be a maximum power setting for this tip; make sure you do not exceed this setting. The conditions are not specified here because they vary depending on the particular sonicator used – most likely you will use a low setting because of the small volume being sonicated. Rinse the tip with distilled water. Immerse the tip into the bacterial solution, but do not let the tip come into contact with the sides of the tube. Sonicate the ice-cold solution for 15 seconds and return the tube to the ice. Sonication will heat up the solution and it is important to sonicate for short periods of time. Perform a total of three cycles of sonications, returning the tube to ice in between sonications. This fraction represents the crude homogenate.

Sonication helps to release the fusion protein from the bacterial cells and reduces the viscosity of the solution by shearing chromosomal DNA.

Removing Insoluble Debris from the Crude Homogenate

1. Spin the crude homogenate at 13,000 rpm in a refrigerated microcentrifuge or in a microcentrifuge kept in a cold box (4°C) for 15 minutes to remove insoluble debris. The insoluble debris represents cells that were not thoroughly sonicated and bacterial membranes. If the pellet is very large, resuspend the pellet, sonicate for additional time and repeat the centrifugation step.
2. Transfer all of the supernatant to a fresh microcentrifuge tube and centrifuge at 13,000 rpm in a refrigerated centrifuge or in a microcentrifuge kept in a cold box (4°C) for 2 minutes. It is critical to ensure that the sample that you load onto the affinity column is free of particulate debris. Insoluble debris from your cell pellet loaded onto your affinity column can make the column flow very slowly. *This cleared crude homogenate will be referred to as the **cell lysate***.
3. Transfer the cell lysate into a 15 ml conical tube and add 3 ml of GST Bind/Wash buffer. Diluting your sample before adding it to the column will help to increase the binding of your sample to the column.
4. Record the approximate volume of the cell lysate (after dilution): ___
5. Remove a 10 μl aliquot of the cell lysate for analysis by SDS-PAGE. Add 10 μl of 2× SDS-PAGE sample buffer to the sample and store at −20°C.
6. Transfer an additional 70 μl of the cell lysate (for fluorescence analysis) to a microfuge tube and store at −20°C.

Purifying Protein by Affinity Chromatography

The affinity resin and buffers should be allowed to equilibrate to room temperature before use.
1. Your instructor will provide you with a slurry of GST-Bind Resin (affinity resin) in a sealed 15 ml test tube. You should have approximately 0.75 ml of settled bead volume. Note: Be gentle! The beads are fragile.

2. Assemble the chromatography column. Attach the stopcock to the bottom of the column. Mount the assembled column on a ring stand.

3. Make sure the stopcock is closed. Invert the GST-Bind Resin to gently mix. Add the uniform slurry of affinity resin material to the column. Open the stopcock to allow the buffer to drip through; do not allow the buffer to drop below the top of the settling beads. Close the stopcock once the buffer level is just above the affinity resin. Also, make sure your column does not leak. If it does, you may have to replace the stopcock.

4. Wash the column with 5 ml of GST Bind/Wash Buffer. Keeping the stopcock wide open, allow the liquid level to drop until just above the bed of resin. Close the stopcock.

5. Add your cell lysate to the column and open the stopcock. Collect the flow-through fraction in a 15 ml conical tube labeled "FT." Close the stopcock once the buffer level is just above the settled resin. The majority of *E. coli* proteins will not bind to the glutathione affinity resin and thus will be found in this flow-through fraction.

    In affinity chromatography, the binding of the fusion protein to the ligand requires a slow flow rate. Allowing the sample to flow through the column by gravity flow achieves this. Forcing the sample through the column at a more rapid rate will decrease the binding efficiency of the target molecule to the ligand.

6. Reapply the flow-through to the column, collecting the flow-through in the same 15 ml tube. Store the flow-through on ice when done. Passing your sample over the column a second time will increase the amount of GST::EGFP fusion protein that binds to the affinity resin, and therefore will increase your yield.

7. Remove a 10 μl aliquot of flow-through sample for analysis by SDS-PAGE. Add 10 μl of 2× SDS-PAGE sample buffer to the sample and store at −20°C.

8. Transfer an additional 50 μl of the flow-through (for fluorescence analysis) to a microfuge tube and store at −20°C.

9. Add 5 ml GST Bind/Wash Buffer to the column, open the stopcock and collect the wash fractions, 1 ml per tube labeled W1–W5 and freeze. Close the stopcock once the buffer level is just above the settled beads.

    In the first wash fraction, you will actually be collecting the remainder of the flow-through fraction; hence this fraction will have a substantial amount of total protein. In the rest of the washes, you will likely have little total protein present. You hope to have little GST::EGFP fusion protein in either the flow-through or washes, as it should be bound to the column at this stage.

10. Wash the column with an additional 5 ml of GST Bind/Wash Buffer to remove any non-GST proteins left behind. Collect a small aliquot of the end of this wash (~100 μl) in a microfuge tube for fluorescence analysis (label as last wash). This will be used to make sure there is no GST::EGFP coming off the column before elution. Close the stopcock once the buffer level is just above the settled beads. Discard the flow-through solution.

11. Label six microcentrifuge tubes for the eluted samples (eluates) as E1–E6.

12. Add 3 ml of GST elution buffer to the column. Collect ~0.5 ml fractions in each of your labeled tubes.

    Reduced glutathione is the competitive agent in the elution buffer that displaces your GST::EGFP fusion protein from the affinity column.

13. Place the tubes with your washes and elution on the UV transilluminator in a dimly lit room. Be sure to wear a UV-protective face shield. Record the fractions which have green fluorescence in your notebook.

14. Store your fractions (crude homogenate, flow-through, washes, elutions) at −20°C for further analysis.

Discussion Questions

1. What is likely to happen if the column capacity of the column were too high? Too low?

2. You accidentally spill your GST-Bind Resin as you are attempting to add the beads to the column. Unfortunately, there is no extra GST-Bind Resin anywhere to be found! What other methods of affinity chromatography could you use to purify your fusion protein?

LAB SESSION 14

Analysis of Purification Fractions

This week you will analyze your purification fractions by SDS-PAGE and quantify the amount of purified GST::EGFP fusion protein using a fluorescence assay. You will also make a fresh replica plate of one of your positive clones for use next week.

Molecular Biology Techniques. DOI: 10.1016/B978-0-12-385544-2.00014-4

LAB SESSION 14A

Analysis of Purification Fractions

Goal: Today you will analyze your purification fractions by SDS-PAGE and quantify the amount of purified GST::EGFP fusion protein using a fluorescence assay. SDS-PAGE will allow you to estimate the purity of the eluted fractions and determine whether any of the recombinant fusion protein was lost during the wash steps. The fluorescence assay will allow you to quantify how much GST::EGFP fusion protein you purified, as well as reveal how much fusion protein was lost in wash steps.

Introduction

The purpose of this laboratory session is to quantify how much GST::EGFP fusion protein you purified, as well as to determine the purity of your elution samples. You will also examine your fractions from affinity purification to see if your fusion protein is present in your unbound or wash fractions. These experiments will use two complementary methods: SDS-PAGE followed by total protein staining and fluorescence detection.

In Lab Sessions 11 and 12, you performed SDS-PAGE on lysates from your transformants and then transferred the proteins to a nitrocellulose membrane and probed with an anti-EGFP antibody in order to confirm the presence of the GST::EGFP fusion protein. In that particular case, we were not concerned with the presence or absence of any other protein(s) in the sample; we simply wanted to confirm that GST::EGFP was expressed. The purpose of the SDS-PAGE we will perform in this laboratory session is to determine which fractions contain fusion protein and to determine whether the fusion protein is free of contaminating protein in the eluted fractions. We will not use antibodies because we need to visualize all of the proteins in the wash and eluted fractions. Instead, we will use GelCode Blue reagent, a G-250 colloidal coomassie-based dye that binds to all proteins in the gel. The cell lysate includes all cellular proteins, including the fusion protein. Ideally, the flow-through and washes will contain all cellular proteins EXCEPT the fusion protein, and the eluate fractions will contain ONLY the fusion protein. The presence of fusion protein in flow-through and/or wash fractions indicates that the column did not have enough affinity resin or that the flow rate was too fast for the fusion protein to bind to the resin. The presence of additional protein bands in the eluate fractions can result from a number of possibilities: incomplete washing of the column; proteolysis of the fusion protein; non-specific binding of proteins to the column or co-purification of endogenous proteins from *E. coli* associated with the fusion protein.

To determine the amount of GST::EGFP protein you purified, you will measure EGFP fluorescence in your samples obtained during purification.

Additionally, by analyzing the amount of fluorescence in your other fractions, i.e., your flow-through and washes, you will be able to quantitatively determine how much of your initial starting material you lost during the purification protocol. As discussed previously, the EGFP protein will fluoresce upon excitation by light at specific wavelengths. While EGFP can be minimally excited by UV light, as you did in Laboratory Sessions 9B and 13b, it is optimally excited by light at 488 nm (blue light). Upon excitation, EGFP emits light at 507 nm, which corresponds to the green color you observed when visualizing your colonies and purification fractions. To quantify the fluorescence of EGFP in your samples, you will use a fluorescence microplate reader. Although your fusion protein is actually GST::EGFP, the fluorescent properties of EGFP should be identical to that of EGFP alone.

The microplate reader is equipped with filters that allow only light of a certain wavelength to pass through. In your experiment, the excitation filter is a 485/20 bandpass filter. This means that light ranging between 465 and 505 nm will go through the filter, while the rest will be blocked. This light will serve to excite the EGFP present in your sample. Upon excitation, the EGFP will fluoresce and the light emitted will pass through the emission filter before being detected by the instrument. The emission filter is a 528/20 bandpass filter, thus allowing light from 508 to 548 nm to pass through. As you can see in comparing the filters, light from the excitation will be filtered out by the emission filter, thereby ensuring that the light detected by the plate reader corresponds specifically to fluorescence from your sample.

Determining protein concentration using fluorescence is similar to a number of other methods for calculating total protein concentration, such as Bradford and the BCA (bicinchoninic acid) assay. In this case though, instead of measuring total protein, you will be measuring only the amount of the GST::EGFP fusion protein in your sample. To do this, you will first make a standard curve using known amounts of recombinant EGFP. The amount of fluorescence in your unknown samples is then determined by comparison to the standard curve. By combining the fluorescence data with that from SDS-PAGE, you will determine both the relative purity of your sample, as well as your yield of GST::EGFP fusion protein.

Laboratory Exercises

SDS-PAGE of Purified Fusion Protein

SDS-PAGE will provide a qualitative analysis of sample purity.

Retrieve last week's samples from the −20° freezer. Label microcentrifuge tubes and aliquot the following:

1. molecular weight marker premixed with sample buffer (10 μl)
2. cell lysate (20 μl – already has sample buffer)
3. flow-through (20 μl – already has sample buffer)
4. wash #3 (10 μl sample + 10 μl sample buffer)[*]
5. wash #4 (10 μl sample + 10 μl sample buffer)[*]
6. wash #5 (10 μl sample + 10 μl sample buffer)[*]
7. eluate 2 (10 μl sample + 10 μl sample buffer)[†]

8. eluate 3 (10 μl sample + 10 μl sample buffer)†
9. eluate 4 (10 μl sample + 10 μl sample buffer)†
10. eluate 5 (10 μl sample + 10 μl sample buffer)†.

*If any of your washes appeared to have green fluorescence, run those instead.
†If eluate fractions other than these had green fluorescence, run those instead.

Once the aliquots have been made, heat the samples for 10 minutes at 95°C or higher before loading. Use tube tabs to lock caps. (The molecular weight marker only needs to be heated for 1 minute.) This will help to prevent aggregation.

Run the gel at constant current (30 mA per gel) until the dye front is at the bottom of the gel (as described in Lab Session 11).

STAINING THE GEL

CAUTION: Wear gloves to prevent blue staining.

1. Turn off the power supply and remove your gel from the apparatus. Be careful not to squeeze the gel.
2. Gently pull apart the two pieces of plastic. The gel should stick to one plate.
3. Cradle the plate in your hand with the gel side facing up. While holding it over your staining tray, direct a stream of distilled water (from a squirt bottle) under the gel to loosen it. Tilt the plate to allow the gel to drop into a square Petri dish.
4. Place the gel in the dish and rinse three times (5 minutes each) with dH₂O with gentle shaking. Label the tray with your station number and lab day on a piece of tape.
5. Mix GelCode Blue Reagent by inverting gently immediately before using.
6. Add approximately 20 ml of stain reagent (enough to cover the gel, but do NOT fill the tray with it!) and shake dish slowly for an hour.
7. Replace the stain reagent with dH₂O and continue shaking. You can probably see the bands at this point, but this step enhances stain sensitivity and weak protein bands will continue to develop. The water may need to be changed several times for optimal results. The gel can be left in water for as long as necessary: 15 minutes to overnight.
8. Photograph your gel and analyze the results.

Fluorescence Analysis of Affinity Purification

Start this assay while your SDS-PAGE is running.

1. Prepare dilutions of recombinant EGFP for a standard curve as described below.

You will be given a tube with 3.2 μl of a 1 mg/ml recombinant EGFP solution in phosphate buffered saline (PBS). Prepare dilutions of the EGFP protein in three tubes as shown in Table 14.1. Attach a new pipette tip to the micropipette each time to make the dilutions. Be sure to mix each tube well after performing the dilution.

2. Prepare samples as follows. Before adding your samples to the microtiter plate, you will dilute them with PBS. Particularly with your cell lysate

Table 14.1 Recombinant EGFP dilutions

Tube	Dilution	Protein concentration	Microplate abbreviation
1	3.2 µl of stock (1 mg/ml) + 636.8 µl PBS	5.0 µg/ml	ST5
2	175.0 µl from tube 1 + 175.0 µl PBS	2.5 µg/ml	ST2.5
3	125.0 µl from tube 1 + 500.0 µl PBS	1.0 µg/ml	ST1
4	175.0 µl from tube 3 + 175.0 µl PBS	0.5 µg/ml	ST0.5

Table 14.2 Dilution scheme for samples

Sample	Microplate abbreviation	Sample volume	Volume of PBS for dilution
Cell lysate (10-fold dilution)	CL10	35 µl	315 µl
Cell lysate (30-fold dilution)	CL30	12 µl	348 µl
Flow-through	FT	35 µl	315 µl
Last wash	LW	35 µl	315 µl
Eluate 1	E1	35 µl	315 µl
Eluate 2	E2	35 µl	315 µl
Eluate 3	E3	35 µl	315 µl
Eluate 4	E4	35 µl	315 µl
Eluate 5	E5	35 µl	315 µl
Eluate 6	E6	35 µl	315 µl

and some of your elution fractions, failure to dilute the samples will result in fluorescence too high for the microplate reader to detect. For your cell lysate, you will make two different dilutions to help ensure that your reading falls within the standard curve. For the other samples, you will prepare 10-fold dilutions of your samples. Table 14.2 provides the details for the dilutions.

3. Add 100 µl of each: PBS (for background), your EGFP standards (for your standard curve) and your unknown samples from your purification to the 96-well microtiter plate, in triplicate, as illustrated in Table 14.3.
4. Measure the fluorescence in the wells using the microplate reader. Your instructor will set up the instrument to properly read the fluorescence. This set-up will vary between different microplate readers. In general, you should be using filters comparable to the following: excitation 485/20; emission 528/20. Furthermore, fluorescence sensitivity should be scaled, if possible, such that the ST5 wells are at the upper range of the scale.
5. Calculate the protein concentration and total protein amount of each sample.

Table 14.3 Schematic of microtiter plate

	1	2	3	4	5	6	7	8	9	10	11	12
A	PBS	PBS	PBS									
B	ST0.5	ST0.5	ST0.5	ST1	ST1	ST1	ST2.5	ST2.5	ST2.5	ST5	ST5	ST5
C	CL10	CL10	CL10	CL30	CL30	CL30	FT	FT	FT	LW	LW	LW
D	E1	E1	E1	E2	E2	E2	E3	E3	E3	E4	E4	E4
E	E5	E5	E5	E6	E6	E6						
F												
G												
H												

TO CALCULATE THE CONCENTRATION AND AMOUNT OF PROTEIN IN EACH SAMPLE

1. Background subtraction.

Open up your data from the microplate reader in a computer program such as Microsoft Excel and average your three PBS wells. This is your background and this value should be subtracted from all your wells.

2. Making a standard curve.

Graph a standard curve derived from your EGFP standards. Your graph should have the concentration of EGFP on the X-axis and fluorescence reading on the Y-axis. Don't forget to include 0,0 as a point on the graph from your blank!

3. Determine the protein concentration of diluted samples.

Calculate the fluorescence average of each triplicate reading for your unknown samples. If one of the numbers is far off, and you suspect that you made a mistake in pipetting, you may discard that reading.

Use a computer program such as Microsoft Excel to calculate the slope of the standard curve: $\mathbf{m = \Delta Y/\Delta X}$ or **Slope = Δfluorescence/ Δconcentration**.

Then plug in $\mathbf{y = mx + b}$ or **fluorescence = slope × concentration** (b = 0 in this case). The concentration you calculate here is the concentration of the sample in the assay well (C_a). Recall that you diluted your samples for the assay, so to determine the protein concentration in the original sample, further calculations are necessary.

4. Determine the concentration in the original sample.

$$C_o\ (\mu g/ml) = \frac{C_a\ (\mu g/ml)}{d_1}$$

Where C_o is the protein concentration in the original sample ($\mu g/ml$), C_a is the protein concentration in the assay microtiter well and d_1 is the dilution of your sample prior to reading fluorescence.

For example, for your flow-through, last wash and eluate samples, $d_1 = 35/(35 + 315) = 0.1$. For your cell lysate sample, you will determine which dilution factor to use based on which reading falls within your standard curve.

Record concentrations in your laboratory notebook.

5. Calculate total protein.

Total protein = Protein concentration $\times$ Total volume.

You recorded the total volume of the cell lysate in Lab Session 13 and collected ~500 μl fractions of your eluates. (If the volume of your elutions is different, adjust accordingly.)

After calculating your total amount of GST::EGFP fusion protein in your various samples, answer the following questions:

1. What percentage of your total fusion protein did you successfully purify? (Sum of amount of eluates/Total amount in cell lysate $\times$ 100.)
2. Proteins are purified for a number of reasons. Some proteins are used as therapeutic agents, i.e., insulin and growth hormones. Another reason is to sell the proteins commercially for scientific use. The recombinant EGFP protein you used to make your standard curve is an example of a commercially available fusion protein (the recombinant EGFP contains a 6x-His tag). Currently, the price of recombinant EGFP is $255 for 100 μg of pure protein. How valuable was your work in the lab over the semester? Do the math to figure out how much the recombinant protein you purified is worth!

LAB SESSION 14B

Replica Plate Positive Clone

Replica plate one of your positive clones onto LB/kan so you will have a fresh inoculum for next week. Label, tape shut and place inverted in the 37°C incubator. Your instructor will save it in the refrigerator after an overnight incubation.

Discussion Questions

1. After running the SDS-PAGE, why do some of the lanes with eluate not show any bands? Why are we collecting the eluate in fractions, rather than all into one tube?

2. Assume that in addition to the fluorescence quantitation, you also performed a Bradford assay on your purified fractions. Would you anticipate calculating more, less, or the same amount of protein in your elutions using this method? Why?

PART 4

Analysis of mRNA Levels

At this point, you have successfully engineered a recombinant DNA molecule and expressed and purified the recombinant fusion protein, GST::EGFP. The final experiments described in this manual aim to measure the levels of *gst::egfp* messenger ribonucleic acid (mRNA). As described in the Central Dogma of Molecular Biology, "DNA makes RNA makes protein." DNA, whether it's genomic DNA (gDNA) or plasmid DNA, is transcribed into RNA by an enzyme called RNA polymerase. In bacteria like *E. coli*, one RNA polymerase is responsible for transcribing different types of RNA. The most abundant type of RNA is ribosomal RNA (rRNA). This type of RNA is a major component of ribosomes. Transfer RNA (tRNA) is also involved in translation, as it helps to add amino acids to the growing polypeptide chain. Protein-coding genes are transcribed into mRNA, which eventually gets translated into protein. Hence, mRNA serves as a messenger to take information from DNA and deliver it to protein-synthesizing ribosomes. While these three types of RNA were once believed to constitute all of a cell's total RNA, we now know that there are many types of RNA, including microRNAs (miRNAs) and Piwi-interacting RNAs (piR-NAs), and their discoveries have expanded our view on just how diverse RNA functions are.

In your project, the PCR product you cloned into pET-41a is the coding sequence for *egfp*. pET-41a already encodes the coding sequence for *gst*. Therefore, when transcription of your positive clone occurs, *gst::egfp* mRNA will be produced. In this final part of the course, you will induce gene expression (transcription and translation) from your plasmid clone as you did in Part 3. This time, you will not be analyzing protein levels, rather you will be analyzing complementary DNA (cDNA) levels, and by inference, mRNA levels. To accomplish this, you will learn how to purify and quantify

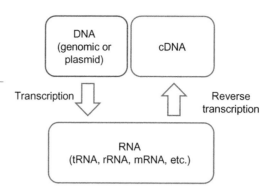

FIG. 15.1
Reverse transcription creates complementary DNA (cDNA). DNA is transcribed into RNA. In a reverse transcription reaction, purified RNA is converted into cDNA. By using random primers in the reverse transcription reaction, all types of RNA can be converted into cDNA.

total RNA (which includes mRNA) from *E. coli* and convert it into cDNA by a reverse transcription reaction (Figure 15.1). In addition, you will be comparing induction of mRNA with IPTG or lactose to no induction at all.

Challenges of Working with RNA

Compared to DNA, RNA is much less stable and should be handled with special care. This difference in stability can be attributed to several features. All RNAs have a 2′ hydroxyl group on the ribose sugar. This hydroxyl group can be involved in a nucleophilic attack of the phosphodiester bond. This results in cleavage of the RNA backbone. DNA contains deoxyribose instead of ribose, so this reaction does not occur in DNA. DNA has smaller grooves in its double helical structure, whereas RNA has larger grooves (and is usually not double-stranded). These larger grooves provide more sequence that can be susceptible to enzymatic cleavage. Finally, RNA is less stable in alkaline (basic) conditions. It should be noted that degradation of bacterial mRNA occurs very quickly, often with half-lives of only several minutes[1].

In addition to these structural characteristics of RNA, the existence of a certain group of enzymes called RNases renders RNA vulnerable to degradation. Microorganisms are sources of RNases that are found in the air, on surfaces, on our skin and in bodily fluids. These ubiquitous enzymes degrade RNA and can be a nuisance to someone trying to isolate and work with intact RNA. In addition, lab techniques that use RNases, such as nuclease protection assays and plasmid purification protocols, may have been previously performed in the same lab, with the same instruments and reagents that you are using. The best solution is prevention: minimize the amount of RNases you expose your samples to. Several steps should be taken when working with RNA in the following lab sessions to accomplish this. First, always wear gloves and change them often. RNases from hands can easily be transferred to pipettes, tubes and reagents. Second, use pipette tips, tubes and water that are certified by the manufacturer as "RNase-free." Autoclaving may be fine to sterilize your tips for general use, but it does not guarantee that RNases have been denatured. Tips that are certified to be "RNase-free" may also contain filters (barriers) that prevent aerosols like RNases from contaminating the pipette shaft. Purchasing sterile tips and water is the easiest way to prevent contamination from these sources. Finally, it is best to set aside pipettes and gel electrophoresis systems that are dedicated solely to RNA work. However, if this is not practical in your lab, you may clean your pipettes and gel electrophoresis systems with a commercially available reagent that inactivates RNAses, such as Ambion's RNaseZAP.[2] You should also use this reagent to clean your bench top before starting a lab session.

Unlike DNA, you do not want to store RNA at 4°C for short-term storage and −20°C for long-term storage. RNA samples are best stored at −80°C. If an RNA sample is to be stored for more than several weeks, adding a salt (sodium acetate) and ethanol solution to the sample before storage at −80°C is recommended. RNA samples should be kept on ice while at your bench; however, each protocol described here will specify if a reaction including RNA should be set up on ice or at room temperature. In general, it is best to work quickly (but carefully) when handling RNA.

The lab sessions in this part of the manual were designed for you to use one of two methods for analyzing mRNA levels. If you have a thermal cycler equipped for quantitative PCR (a real-time system), it is recommended that you perform Lab Sessions 15, 16 and 17. If you do not have a real-time system, you should perform Lab Sessions 15, 18 and 19. Both methods will generate comparable results, although real-time PCR is much more sensitive. Both methods begin with the same steps (induction, total RNA purification, DNase digestion, quantification, and reverse transcription). After that, the PCR and data analysis steps diverge.

References

1. <http://www.qiagen.com/>
2. <http://www.ambion.com/>

LAB SESSION 15

Total RNA Purification

Goal: In this laboratory session, you will inoculate cultures of a positive transformant and then your instructor will induce these cultures with IPTG, lactose, or no inducer. You will then purify total RNA from these cells, treat each RNA sample with DNase and quantify the RNA. Samples will be saved at −80°C until next week.

Molecular Biology Techniques. DOI: 10.1016/B978-0-12-385544-2.00015-6

LAB SESSION 15A

Interim Laboratory Session

Laboratory Exercise

Inoculate Cultures for RNA Purification

1. Inoculate a positive clone into 3 ml LB/kan broth in a snap-cap tube. Do this by picking bacterial growth from your most recent master plate with a sterile toothpick and dropping it into the tube containing LB/kan.
2. Label the tube with your lab day and station number.
3. Label three empty snap-cap tubes with your lab day, station number and either "no," "IPTG" or "lactose."
4. Place the inoculated culture and the three empty, labeled tubes in the designated racks. The LB/kan tubes will be refrigerated to retard growth. The night before your lab they will be placed in a shaker incubator at 37°C. After growing overnight, the cultures will be ready for induction. The empty, labeled tubes will be used for this induction by your instructor.

LAB SESSION 15B

Total RNA Purification

Introduction

Recall that in Part 3, IPTG was added to your bacterial cultures (liquid broth or agar plate) to induce gene expression. This induction was measured at the protein level (visualization of GST::EGFP by UV light, SDS-PAGE and western blot). Gene expression refers to both transcription and translation, with the end result being protein production. To analyze induction at the mRNA level, cultures will similarly be induced with IPTG to derepress the lac operator of pET-41a. IPTG is a non-hydrolyzable analog of lactose, so you will try inducing with IPTG and compare that to induction with the same concentration of lactose. These two conditions will serve as your experimental conditions and no inducer at all will serve as your control.

After induction, total RNA will be purified from the *E. coli* cells. Total RNA refers to all the RNA content of a cell including, but not limited to, mRNA, rRNA and tRNA. The most abundant type of RNA is rRNA and the type of RNA we are interested in analyzing, mRNA, makes up a very small fraction of total RNA. We will use the Qiagen RNeasy Mini Kit to purify total RNA (Figure 15.2). First, *E. coli* cells are incubated with lysozyme to degrade the bacterial cell wall. Then, a lysis buffer is added that contains guanidine thiocyanate. This chemical is a denaturant that quickly inactivates RNases. Next, the lysate is vortexed and passed through a column called a QIAshredder homogenizer. These homogenization steps will serve to reduce viscosity (large molecular weight genomic DNA is sheared) and remove insoluble material from the bacterial lysate. The QIAshredder is quicker and easier than homogenization methods involving beads or a needle and syringe. Next, ethanol is added to create conditions favorable for RNA binding to the RNeasy column membrane. The final purification steps are similar to the silica adsorption steps used in this manual to purify plasmid DNA. The sample is applied to an RNeasy column, contaminants are washed away and total RNA is eluted in RNase-free water. The column membrane will selectively bind RNA larger than 200 nucleotides; therefore the eluted RNA does not include tRNA or other low molecular weight RNA. This is beneficial in our case, since we are interested in analyzing mRNA.

The RNeasy columns will remove most of the DNA found in the lysates; however, for downstream steps, efficient DNA removal is required. Even trace amounts of either *E. coli* genomic DNA (found naturally in all *E. coli* cells) or plasmid DNA (transformed into *E. coli* cells) can be detected with a technique as sensitive as RT-PCR. The template that you are interested in amplifying in the PCR step is cDNA directly made from mRNA (Lab Sessions 16 and 18). Therefore, to achieve accurate quantification

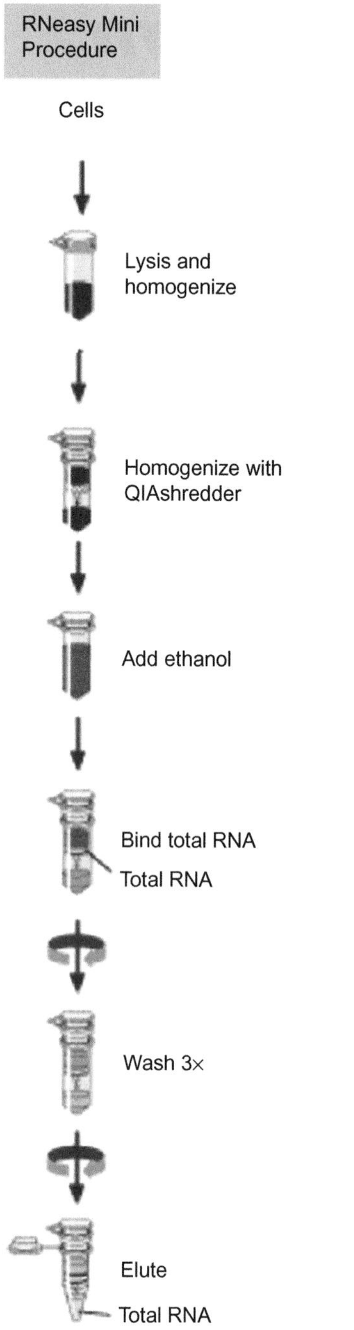

RNeasy Mini
Procedure

Cells

Lysis and
homogenize

Homogenize with
QIAshredder

Add ethanol

Bind total RNA

Total RNA

Wash 3×

Elute

Total RNA

FIG. 15.2
RNeasy Mini Kit flowchart. Copyright 2010
Qiagen Corporation. Used with permission.

128

of mRNA levels, you must eliminate any DNA in the RNA sample that may serve as template. Although DNase digestion can be performed "on-column," it is more efficient to perform the DNase treatment in solution on the eluted RNA. Finally, DNase will be inactivated and salts removed by a final purification step on another RNeasy mini column. This step is analogous to the column purification of your linearized plasmid DNA.

It is often valuable to assess the integrity (lack of degradation) of purified RNA before proceeding. In methods such as northern blotting or micro-array sample preparation, intact RNA is vital for successful experiments and subsequent analysis. One way that RNA integrity can be analyzed is by agarose gel electrophoresis. Due to RNA's single-strandedness, it has the ability to fold back on itself and form secondary structures through complementary base pairing. To prevent this secondary structure from affecting migration through an agarose gel, denaturing conditions are used. Rather than a native agarose gel like those you prepared for analyzing DNA, denaturing agarose gels (usually containing formaldehyde) are used. RNA samples are also heated in a loading dye that contains the denaturant formamide before loading them in the gel. RNA is not stable in alkaline conditions, so a neutral buffer is used in both the gel and running buffer. MOPS (3-[N-Morpholino]-propanesulfonic acid) buffer is most commonly used for electrophoresis of RNA. These steps ensure that RNAs are kept denatured (single-stranded) and migrate based on size alone. RNA molecular weight markers are run next to RNA samples, since fragments of a DNA molecular weight marker would migrate differently than RNA fragments of the same size due to their double-strandedness.

Results of a typical denaturing agarose gel loaded with total RNA purified from *E. coli* are shown in Figure 15.3. Remember that total RNA was purified, so only the most abundant types of RNA will be detected using ethidium bromide or a similar fluorescent nucleic acid dye like GelRed. That is why mainly rRNA bands are visible on the gel. In *E. coli*, 23S rRNA and 16S rRNA are clearly seen migrating at 2.9 kb and 1.5 kb, respectively. Notice in lane 2 that there is approximately twice as much 23S rRNA as 16S rRNA. This 2:1 ratio is one indication of intact RNA. There are also sharp bands seen in lane 2, demonstrating little degradation. If RNA is degraded, smearing will be evident. Lane 3 shows RNA that has been treated with DNase. Three differences from the RNA without DNase digestion can be seen. First, there is slight smearing in lane 3 below each rRNA band, indicating some degradation. If the RNA was heavily degraded, no distinct bands would be apparent at all, and the entire lane would be a smear. Second, the 2:1 ratio is not maintained. Lane 3 looks as if 23S and 16S rRNA are equal in abundance. Third, notice that the rRNA bands in lane 3 have migrated further than the rRNA bands in lane 2. This means that the rRNAs are smaller than normal, consistent with degradation. These results are not surprising, since the DNase treatment and subsequent purification are additional physical manipulations to the sample. Recall that RNA, especially bacterial RNA, can degrade rapidly. For our RT-PCR experiment, we will include the DNase treatment but skip the denaturing agarose gel electrophoresis. Even though the amounts of 23S and 16S rRNA are apparently equal, this is not the only indicator of RNA integrity. Furthermore, the amount of RNA degradation that seems to have occurred due to the DNase treatment steps is still tolerable and will not adversely affect the

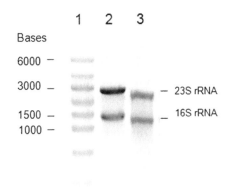

Bases

6000 —
3000 —
1500 —
1000 —

— 23S rRNA
— 16S rRNA

FIG. 15.3

Denaturing 1% agarose gel electrophoresis
of total RNA from *E. coli*. Lane 1: Fermentas
RiboRuler High-Range RNA Ladder. Lane 2: 1 µg
total RNA without DNase digestion. Lane 3: 1 µg
total RNA with DNase digestion.

reverse transcription and PCR steps. The need for eliminating contaminating DNA in this experiment outweighs any minor degradation the RNA has undergone.

Finally, RNA will be quantified by a spectrophotometer (or NanoDrop). Like DNA, RNA also absorbs UV light maximally at 260 nm. However, RNA has a different average extinction coefficient than double-stranded DNA due to its single-strandedness. An absorbance of one unit at 260 nm corresponds to approximately 40 µg single-stranded RNA/ml. When calculating the concentration of an RNA sample, the correct extinction coefficient must be used. The purity of an RNA sample can also be estimated by a spectrophotometer. Recall that pure DNA should have an A_{260}/A_{280} ratio of approximately 1.8. Pure RNA should have an A_{260}/A_{280} ratio of approximately 2.0. Anything less than 2.0 could indicate possible protein contamination; however, the amount of contaminantion would have to be extreme to significantly affect this ratio.[1] As both DNA and RNA absorb UV light at 260 nm, an acceptable A_{260}/A_{280} ratio cannot rule out DNA contamination. Remember that we are including a DNase digestion to minimize the amount of contaminating DNA. Ratios of 1.9–2.1 are generally acceptable for RNA that has been treated with DNase.

Laboratory Exercises

Purification of Total RNA

We will purify total RNA from bacteria using the QIAGEN RNeasy Mini Kit. QIAshredder spin columns can be purchased separately, or obtained as part of the QIAGEN RNeasy Plant Mini Kit. A lysozyme solution is also required that is not supplied by the Mini Kit or Plant Mini Kit. Another option is to use the RNeasy® Protect Bacteria Mini Kit.[2] The following protocols have been modified from the manufacturer's handbook (QIAGEN RNeasy® Mini Handbook 09/2010).[3]

SAMPLE PREPARATION

Steps 1–3 will be done for you.

1. Pipette 1 ml of the overnight culture into each of three empty, labeled tubes.
2. Pipette 2 ml of the appropriate 2× YT broth into each tube: 2× YT/kan/**IPTG**, 2× YT/kan/**lac**, or 2× YT/kan (**no inducer**).
3. Incubate in the shaker at 37°C for 2 hours (do not allow the incubation to proceed any longer than 2 hours).
4. Pipette 1.5 ml of each *E. coli* culture into three labeled microcentrifuge tubes.

IMPORTANT: Clearly label all tubes and spin columns throughout this protocol to be sure your control (no inducer) and experimental samples (IPTG and lactose) are kept separate.

5. Harvest the cells by centrifugation at 5000 rpm for 5 minutes in a refrigerated microcentrifuge (4°C). Make sure this centrifuge is pre-chilled. This is the only step performed at 4°C. All subsequent steps, including centrifugations, are performed at room temperature.

6. While the cells are spinning, dilute the lysozyme stock 1:100 by pipetting 5 µl into a new microcentrifuge tube and adding 495 µl of water. Keep on ice.

7. Carefully remove (decant) all supernatant into an appropriate bacterial liquid waste container and invert the tube over a paper towel to be sure no residual liquid medium remains. Incomplete removal of the medium will inhibit lysis.

8. Resuspend each bacterial pellet in 100 µl of diluted lysozyme from step 6. Pipette up and down until no clumps remain.

9. Allow the cells to lyse at room temperature for 5 minutes. During this time, set your heat block to 37°C.

10. Pipette 350 µl Buffer RLT into each tube and vortex vigorously for 1 minute.

11. Transfer the lysate to a QIAshredder spin column (lilac) placed in a 2 ml collection tube and centrifuge for 2 minutes at full speed. IMPORTANT: do not throw away the flow-through. You will need it in the next step!

12. Carefully transfer the **supernatant of the flow-through** to a new microcentrifuge tube without disturbing the cell-debris pellet in the collection tube. Use only this supernatant in subsequent steps.

13. Add 250 µl of ethanol (96–100%) to the saved supernatant and mix immediately by pipetting. Do not centrifuge. Proceed immediately to step 14.

14. Transfer the sample (usually 700 µl), including any precipitate that may have formed, to an RNeasy spin column (pink) placed in a 2 ml collection tube. Close the lid gently and centrifuge for 15 seconds at $\geq 8000 \times g$ ($\geq 10,000$ rpm). Discard the flow-through.[*] Reuse the collection tube in step 15.

15. Add 700 µl Buffer RW1 to the RNeasy spin column. Close the lid gently, and centrifuge for 15 seconds at $\geq 8000 \times g$ ($\geq 10,000$ rpm) to wash the spin column membrane. Discard the flow-through.[†] Reuse the collection tube in step 16.

16. Add 500 µl Buffer RPE to the RNeasy spin column. Close the lid gently and centrifuge for 15 seconds at $\geq 8000 \times g$ ($\geq 10,000$ rpm) to wash the spin column membrane. Discard the flow-through. Reuse the collection tube in step 17.

17. Add 500 µl Buffer RPE to the RNeasy spin column. Close the lid gently and centrifuge for 2 minutes at $\geq 8000 \times g$ ($\geq 10,000$ rpm) to wash the spin column membrane. The long centrifugation dries the spin column membrane, ensuring that no ethanol is carried over during RNA elution. Residual ethanol may interfere with downstream reactions.

18. Place the RNeasy spin column in a new, labeled microcentrifuge tube. Add 50 µl RNase-free water directly to the center of the spin column membrane. Close the lid gently and centrifuge for 1 minute at $\geq 8000 \times g$ ($\geq 10,000$ rpm) to elute the RNA.

---

[*] Flow-through contains Buffer RLT and is therefore not compatible with bleach. Dispose of it according to your institution's waste-disposal guidelines.
[†] Flow-through contains Buffer RW1 and is therefore not compatible with bleach. Dispose of it according to your institution's waste-disposal guidelines.

DNase Digestion

Do not vortex the reconstituted DNase I. DNase I is especially sensitive to physical denaturation. Mixing should only be carried out by gently inverting the tube.

1. To each tube of eluted RNA (50 µl), add the following for a total volume of 100µl:
   - 30 µl RNase-free water
   - 10 µl Buffer RDD
   - 10 µl DNase.
2. Mix by flicking and place in the 37°C heat block for 30 minutes.
3. Add 350 µl Buffer RLT and mix well by pipetting.
4. Add 250 µl ethanol (96–100%) to the diluted RNA and mix well by pipetting. Do not centrifuge. Proceed immediately to step 5.
5. Transfer the sample (700 µl) to an RNeasy Mini spin column placed in a 2 ml collection tube. Close the lid gently and centrifuge for 15 seconds at $\geq 8000 \times g$ ($\geq 10,000$ rpm). Discard the flow-through.
6. Add 500 µl Buffer RPE to the RNeasy spin column. Close the lid gently and centrifuge for 15 seconds at $\geq 8000 \times g$ ($\geq 10,000$ rpm) to wash the spin column membrane. Discard the flow-through. Reuse the collection tube in step 7.
7. Add 500 µl Buffer RPE to the RNeasy spin column. Close the lid gently and centrifuge for 2 minutes at $\geq 8000 \times g$ ($\geq 10,000$ rpm) to wash the spin column membrane.
8. Place the RNeasy spin column in a new, labeled microcentrifuge tube. Label each tube with your lab day and station number and the inducer used for that sample (no, IPTG, or lac). Add 50µl of RNase-free water directly to the center of the spin column membrane. Close the lid gently and centrifuge for 1 minute at $\geq 8000 \times g$ ($\geq 10,000$ rpm) to elute the RNA.

Quantification of RNA

Depending on the equipment available in your lab, you will use a spectrophotometer or NanoDrop to quantify each of your DNase-digested RNA samples. In addition, you will assess each sample's purity.

1. Refer to Lab Session 2B for details on how to use either instrument.
2. When blanking the instrument, use the same RNase-free water that was used to elute your samples.
3. IMPORTANT: If using the NanoDrop, you must set the software to read RNA by choosing "RNA-40" from the drop-down list. RNA has a different extinction coefficient than DNA, so the concentration of your sample will be inaccurate if the software uses DNA's extinction coefficient in calculations. If you are calculating concentration manually from a spectrophotometer reading, use RNA's extinction coefficient (40 µg/ml) in your equation.
4. Record the concentration of each sample in your notebook and write it on the side of the tube.
5. Calculate and record the 260/280 ratio of each sample. If the ratio is not in the 1.9–2.1 range, let your instructor know.
6. Save samples in a −80°C freezer until your next lab session.

Reminder: If you are going to perform quantitative PCR with a real-time-capable thermal cycler, proceed to Lab Session 16 next week. If you are going to perform semi-quantitative PCR with a standard thermal cycler, proceed to Lab Session 18 next week.

References

1. Glasel JA. 1995. Validity of nucleic acid purities monitored by a260/a280 absorbance ratios. *Biotechniques*. 18:62–63.
2. QIAGEN RNAprotect® Bacteria Reagent Handbook. December 2005.
3. QIAGEN RNeasy® Mini Handbook. September 2010.

Discussion Questions

1. Could we quantify just the amount of mRNA in our samples by using a spectrophotometer or NanoDrop? Why?
2. Will the 260/280 ratio differ between intact RNA samples and degraded RNA samples? Why?

LAB SESSION 16

Analysis of *gst::egfp* mRNA Levels by RT-qPCR: Part 1

Goal: Total RNA that you purified in the last session will be converted to cDNA by a reverse transcription reaction. Then, that cDNA will be used as a template for quantitative (real-time) PCR. You will amplify DNA with primers specific to *egfp* (your target) and *23S* rRNA (your reference gene).

Introduction

Several techniques have been developed to detect RNA and analyze its abundance. Northern blots are similar to the western blots that you performed, but an RNA sample is loaded on an agarose gel. The samples are then transferred to a nylon membrane and incubated with a DNA or RNA probe specific to a gene of interest. The intensity of bands detected on the membrane is used to quantify RNA levels. This technique, although still used, is labor-intensive and requires a relatively large amount of fully intact RNA. A type of nuclease protection assay called an RNase protection assay (RPA) also relies on a labeled probe to detect an RNA that it has hybridized to. This method is also laborious, although it is more sensitive than a northern blot. In contrast, newer methods based on PCR (which you also performed in this course) are less time-consuming, more sensitive, and are able to tolerate partially degraded RNA samples. In Lab Sessions 16 and 18, **reverse transcription PCR (RT-PCR)** will be performed. The first part of this method is reverse transcription (also called cDNA synthesis). This is the process of converting RNA to DNA using an enzyme called reverse transcriptase (see Figure 15.1). The resulting DNA is complementary to the RNA template, so it is called complementary DNA (cDNA). In the second part of this method, the cDNA is used as a template in PCR. In this way, the amount of PCR product is indicative of the starting amount of input cDNA (and by inference, RNA that it was generated from). RT-PCR can be performed in two steps, with an RT reaction occurring first; then using that product, as a template in a PCR reaction. Alternatively, RT-PCR can be performed in one step, with both reactions occurring in one tube. For the following lab sessions, we will use a two-step protocol.

Lab Sessions 16 and 17 use a highly-sensitive version of RT-PCR called **quantitative RT-PCR (RT-qPCR)**. A few clarifications concerning nomenclature should be made at this point. Although other acronyms can be found for this method, RT-qPCR (with q in front of PCR) will be used in this manual because it is the PCR step that is quantitative. The RT stands for reverse transcription. Quantitative PCR can also be performed on DNA as the starting material, so there is no need for reverse transcription in that case. This is simply abbreviated qPCR. Both types of qPCR mentioned here can also be referred to as real-time PCR because the abundance of a product is measured in real-time, as the products are being amplified. Lab Sessions 18 and 19 are included as an option for a less sensitive version of RT-PCR called semi-quantitative RT-PCR.

Reverse Transcription

What is required for a reverse transcription reaction? The enzyme that catalyzes this reaction is a reverse transcriptase, and there are several commercially available enzymes usually isolated from viruses called retroviruses. These viruses have an RNA, rather than a DNA, genome so they must encode a reverse transcriptase to convert their genome to DNA. Human Immunodeficiency Virus (HIV) is an example of a retrovirus. The reverse transcriptase that we will use is isolated from Maloney Murine Leukemia Virus (MMLV). Along with its supplied buffer, primers are necessary to prime cDNA synthesis. Instead of forward and reverse primers such as those that are used in PCR, reverse transcription reactions use only one primer, because only one strand of cDNA is made. Primers fall into three categories: randomers (ours are hexamers), oligo-dT and gene-specific. Oligo-dT primers will specifically anneal to polyA tails found on most eukaryotic mRNAs. We cannot use this type of primer since we are using prokaryotic RNA. We want to prime cDNA synthesis from all the RNAs in the cell, so gene-specific primers will not be suitable in this case. We will use random hexamers that have the ability to anneal to all types of RNA, without knowledge of sequence. Random hexamers are a pool of primers designed to represent all possible combinations of six-base-pair stretches. This design allows for the primers to bind to all RNA sequences. Finally, dNTPs (like those used in PCR) will be included in the reverse transcription reactions. These will be incorporated into the newly synthesized cDNA strand.

Whenever RT-PCR is performed, an important aspect to control for is contaminating DNA. As mentioned above, RT-PCR uses RNA as starting material. The goal is to synthesize cDNA from that RNA, followed by amplification of the cDNA with gene-specific primers. However, if DNA is still present in the RNA sample, it can be amplified with the gene-specific primers as well, leading to inaccurate quantification of mRNA levels. To control for any contaminating DNA remaining in your samples even after DNase treatment, negative control reactions should be included that do not include reverse transcriptase. Without this enzyme present, the only DNA in the sample will be contaminating DNA (either genomic or plasmid) which will result in a PCR product. However, if you have effectively eliminated DNA in your starting material, amplification using an "−RT" cDNA template should show no product. The corresponding "+RT" cDNA template should result in a product.

Quantitative PCR

Quantitative PCR does not rely on visualization using gel electrophoresis. Instead, quantification of amplified DNA is achieved using fluorescence measurements. Rather than examining fluorescence at the end of the PCR experiment, fluorescence is monitored as it is produced (in "real-time"). There are two main types of detection chemistries that can accomplish this. A fluorescent dye called SYBR® Green binds to double-stranded DNA and, upon excitation by a light source, fluoresces approximately 1000 times brighter than when it is in its unbound form. Therefore, the amount of fluorescence in a reaction is proportional to the amount of double-stranded DNA. The second type of detection chemistry commonly used in qPCR is a 5′ nuclease assay called a TaqMan® assay. This involves not only gene-specific primers, but also a gene-specific DNA probe. The TaqMan® probe is designed to anneal to the target gene (cDNA) somewhere between the forward and reverse primers. This probe contains a reporter dye, most commonly 6-carboxyfluorescein (6-FAM™) at the 5′ end and a non-fluorescent quencher at the 3′ end. The close proximity of the reporter to the quencher prevents fluorescence. However, when PCR amplification occurs, DNA polymerase displaces the bound probe from the template. The 5′ exonuclease activity of the DNA polymerase cleaves the probe, resulting in fluorescence proportional to the amount of product formed. Although the TaqMan assay is more sensitive and eliminates formation of non-specific products, it requires a probe for each gene you want to quantify. SYBR® Green has the disadvantage of binding primer dimers and non-specific products as well as specific products (because it cannot discriminate between different double-stranded DNAs) but it is more cost-effective, especially if you are assaying several different genes. We will use SYBR® Green in this manual as a way to quantify mRNA levels.[1]

RT-qPCR is a powerful, sensitive technique for analyzing gene expression, but designing a successful RT-qPCR experiment requires attention to many details. First, since we wish to quantify the level of *gst::egfp* mRNA (our target gene), we need primers that will specifically amplify this sequence from cDNA. In general, it is suggested that primers used in qPCR follow these guidelines:

- amplify a sequence (amplicon) between 50 and 150 bp;
- GC content should be 20–80%;
- avoid runs of the same nucleotide (runs of four or more guanidines, especially);
- T_m should be close to 60°C;
- T_ms of forward and reverse primers should not differ by more than 1°C.

Primer-design software can be useful when default settings are changed to specify the guidelines outlined above. In addition to target gene primers, a second gene will be assayed. This provides an internal control that will be used to normalize experimental results. It is referred to as a reference gene. The reference gene used here is endogenous *E. coli 23S* rRNA. Reference genes must be expressed in each sample, but, in order to be considered valid, expression levels cannot vary. The only gene we expect

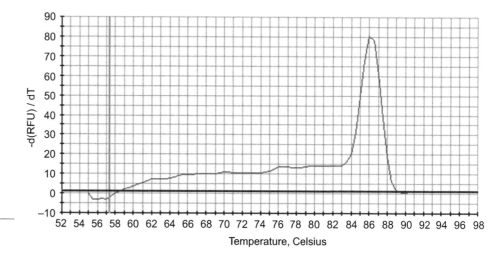

FIG. 16.1

Melt curve analysis of 23S rRNA amplicon.

to change with treatment (induction with IPTG or lactose) is *gst::egfp*. Due to the limitations of SYBR® Green mentioned above, it is imperative that qPCR primers only produce one specific amplicon. Remember that we will not be visualizing the PCR results on an agarose gel. We will be relying on fluorescence produced during the reaction to indicate the amount of DNA present. If fluorescence can be attributed to some amplification of primer-dimers or other non-specific products, we will not have an accurate quantification of *gst::egfp*. The primers you will use have been analyzed to verify that only one amplicon is produced. The most common way to accomplish this is through a melt curve analysis. A melt curve (also called a dissociation curve) is a graph that displays information on how DNA dissociates (melts) upon heating. This analysis can be performed at the end of the qPCR program by gradually increasing the temperature in the thermal cycler. The changes in fluorescence are plotted against temperature. An example of a melt curve is shown in Figure 16.1. A single, narrow peak indicates one specific amplicon was formed. The *23S* rRNA amplicon shown has a T_m of approximately 86°C. Due to their shorter length, primer-dimers, if present, will produce a peak with a T_m lower than that of the desired amplicon.

To produce truly quantitative RT-PCR results, both biological and technical replicates are required. In our experiment, biological replicates would consist of RNA samples from independently induced bacterial cultures. Technical replicates would consist of multiple PCR reactions, each with the same cDNA template. However, so that the entire class's samples will fit in one thermal cycler, we will not perform replicate experiments.

The main disadvantage of RT-qPCR is expense, because it requires a thermal cycler capable of detecting fluorescence. However, if a capable thermal cycler is available to use, it has many advantages over semi-quantitative RT-PCR in gene expression studies. It is a relatively short procedure, since no gel electrophoresis is required. Use of a sensitive fluorescent dye (much more so than ethidium bromide or GelRed) enables detection of genes expressed even at very low levels. In addition, RT-qPCR has a broader dynamic range of detection (the range of input RNA that the assay can detect). These favorable qualities make qPCR a powerful method

Table 16.1 Reverse transcription master mixes

Reagent	For 1 RT reaction	+RT master mix	−RT master mix
1. 5× iScript select reaction mix	4 µl	16 µl	16 µl
2. Random primer	2 µl	8 µl	8 µl
3. iScript reverse transcriptase	1 µl	4 µl	0 µl
4. nuclease-free water	0 µl	0 µl	4 µl
Total	7 µl	28 µl	28 µl

in gene expression studies, and it is often used in validating microarray results. Beyond the basic research lab, qPCR is also used in clinical and industrial settings. Determining viral load, detecting pathogens, genotyping, determining drug therapy efficacy and quality control can all be achieved with this technique.

Laboratory Exercises

Reverse Transcription

The following protocol has been adapted from Bio-Rad's iScript™ Select cDNA Synthesis Kit.

1. Remove your three RNA samples from the −80°C freezer and place them immediately on ice to thaw.
2. Label six PCR tubes with the following: −no, −IPTG, −lac, +no, +IPTG, +lac.
3. Calculate the volume of RNA needed for 1 µg based on your quantification from last week. Use the equation:

   Volume needed (μl) = Amount desired (μg)/Concentration (μg/μl)

   Do this for each of the three samples and record the volumes in your notebook.
4. Pipette this volume into the appropriate PCR tube on ice. Each RNA sample will get pipetted twice: once to a "−" tube and once to a "+" tube.
5. Add enough nuclease-free water to each tube for a final volume of 13 µl. This volume may be different for each of the three samples.
6. To ensure that multiple samples are treated equally with each reagent and for ease of pipetting, you will make a master mix of each of the components listed in Table 16.1. One master mix will contain reverse transcriptase (+RT) and the second master mix will just contain nuclease-free water (−RT). Add the following components, in the order shown, to a microcentrifuge tube on ice.
7. Vortex to mix, perform a quick spin and keep on ice.
8. Pipette 7 µl of the correct master mix into each of the three PCR tubes on ice. Pipette up and down to mix. IMPORTANT: Change pipette tips before taking from the master mix again.

9. Mix gently by flicking and keep on ice until all stations are done with this reaction set-up.

10. Place PCR tubes in a thermal cycler programmed as follows:
    - 25°C   5 minutes
    - 42°C   30 minutes
    - 85°C   5 minutes
    - 4°C   ∞.

11. Once 4°C is reached, stop the thermal cycler and place your six PCR tubes on ice.

Quantitative PCR (qPCR)

The cDNA that you synthesized will now be used as a template in qPCR reactions. You will amplify *gst::egfp* cDNA (from the mRNA that was produced from your expression vector) in each of your cDNA samples. The primers specifically anneal to the *egfp* sequence. You will also amplify *E. coli 23S* rRNA (a gene found on the chromosome of all *E. coli* cells). This amplicon will serve as a reference gene (an internal control for different amounts of input RNA in each reaction). Next week, you will quantify the level of *gst::egfp* cDNA (and by inference, mRNA) in your positive clone after either no induction, induction with IPTG, or induction with lactose.

Note to instructor: A thermal cycler with 96 wells will hold PCR reactions for eight groups of students (12 reactions per group). If you have more than eight groups of students, you will need to either use a second thermal cycler or to eliminate reactions in order for one thermal cycler to accommodate all the reactions. Preferably, eliminate the lactose-induced samples to leave 8 reactions instead of 12. The thermal cycler will then accommodate up to 12 groups. Since both *gst::egfp* and *23S* need to be amplified for quantification purposes and −RT reactions ensure contaminating DNA is not contributing to the quantification of mRNA, it is important to include these if at all possible.

The primers used to specifically amplify *gst::egfp* from your positive clone are:

- egfp16 F: CGACGGCAACTACAAGACC
- egfp16 R: GTCCTCCTTGAAGTCGATGC.

The primers used to specifically amplify *23S* rRNA from your positive clone are:

- 23S16 F: GACGGAGAAGGCTATGTTGG
- 23S16 R: GTTGCTTCAGCACCGTAGTG.

The following protocol has been adapted from Bio-Rad's iQ™ SYBR® Green Supermix instructions.

1. The cDNA that you just synthesized will now be used as a template in a quantitative PCR assay. A very small amount of cDNA is needed (the equivalent of 10 pg of RNA), so you will dilute each sample as shown in Table 16.2. Using a P10, pipette the indicated reverse transcription reaction volume into a new microcentrifuge tube labeled "cDNA dilution 1." Change tips to add sterile water. Vortex to mix well. Use a new tip to

Table 16.2 cDNA dilutions

	Reverse transcription reaction	Sterile dH$_2$O
cDNA dilution 1	2 µl original reaction	998 µl
cDNA dilution 2	2 µl dilution 1	98 µl

Table 16.3 qPCR master mixes

Reagent	For 1 PCR reaction	*egfp* master mix	*23S* master mix
1. Sterile dH$_2$O	6.75 µl	47.25 µl	47.25 µl
2. Primer mix	0.75 µl	5.25 µl *egfp* primer mix	5.25 µl *23S* primer mix
3. iQ SYBR® Green Supermix	12.5 µl	87.5 µl	87.5 µl
Total	20 µl	140 µl	140 µl

pipette into a new microcentrifuge tube labeled "cDNA dilution 2." Change tips to add nuclease-free water. Vortex to mix well.

2. Obtain 12 PCR tubes and label them 1–12 on the side of the tube. Do not label the lids. It is a good idea to record in which wells of the thermal cycler your tubes are placed, in case the ink wears off the sides. Also label the side of each tube with your station number. Place these in a PCR tube rack on ice.

3. Label two microcentrifuge tubes "*egfp* master mix" and "*23S* master mix." To ensure that multiple samples are treated equally with each reagent and for ease of pipetting, you will make a master mix of each of the components listed in Table 16.3. Each master mix will contain everything necessary for the PCR reaction except for the cDNA. One master mix will contain the primers specific for *egfp*. The other master mix will contain the primers specific for *23S* rRNA.

Note to instructor: Table 16.3 shows the volumes needed for six PCR reactions with each primer pair (three +RT and three −RT reactions), including an additional 20 µl volume to account for pipetting error. If you decided to use a larger or smaller sample number, you will need to adjust the volumes accordingly.

4. Pipette each reagent in the order shown into a microcentrifuge tube on ice.

5. Vortex well to mix and return to ice.

6. Pipette as accurately as possible 20 µl of the *egfp* master mix into each PCR tube labeled 1–6. Change tips between each tube. Pipette 20 µl of the *23S* master mix into each PCR tube labeled 7–12. Avoid creating bubbles that may interfere with the fluorescence reading.

7. Pipette as accurately as possible 5 µl of the appropriate diluted cDNA into the tubes as shown in Table 16.4. Be sure to take from the tube

Table 16.4 qPCR setup

Tube	1	2	3	4	5	6	7	8	9	10	11	12
1. Master mix (20 µl)	egfp	egfp	egfp	egfp	egfp	egfp	23S	23S	23S	23S	23S	23S
2. cDNA (5 µl)	−no	+no	−IPTG	+IPTG	−lac	+lac	−no	+no	−IPTG	+IPTG	−lac	+lac

Table 16.5 qPCR program

Cycle	Repeats	Step	Dwell Time	Set Point
1	1	1	3 minutes	95°C
2	40	1	15 seconds	95°C
		2*	30 seconds	60°C

labeled "cDNA dilution 2." Also, change tips between samples. Avoid creating bubbles that may interfere with the fluorescence reading.

8. Keep PCR tubes on ice until all stations are complete.
9. Place tubes in a thermal cycler that has been programmed as shown in Table 16.5.

*Note to instructor: Make sure that the thermal cycler is programmed to collect real-time data during Cycle 2, Step 2.

10. The program will take approximately 1.5 hours to complete. The data collected during the course of the experiment will be saved by your instructor and you will analyze it next week.

Discussion Questions

1. If you had primed reverse transcription of your *E. coli* RNA with an oligo-dT primer rather than a random hexamer, how would that affect our experiment?
2. You are amplifying both *gst::egfp* mRNA and *23S* rRNA in the PCR steps. These are two different types of RNA (messenger and ribosomal, respectively). Could an *E. coli* mRNA rather than an rRNA be amplified as a reference gene? Why? What are the main criteria for selecting a reference gene?

Reference

1 Applied Biosystems. Guide to Performing Relative Quantitation of Gene Expression Using Real-Time Quantitative PCR. <http://www3.appliedbiosystems.com/cms/groups/mcb_support/documents/generaldocuments/cms_042380.pdf/>

LAB SESSION 17

Analysis of *gst::egfp* mRNA Levels by RT-qPCR: Part 2

Goal: You will analyze RT-qPCR data obtained in the previous session to determine the relative levels of *gst::egfp* mRNA in *E. coli* induced with IPTG and lactose compared to no inducer.

Introduction

During the qPCR run, fluorescence readings were recorded during the annealing/extension step of every sample at every cycle (in "real-time"). By the end of the run, a complete amplification plot (fluorescence readings versus cycle number) was generated for each sample. Now, you must analyze these plots in order to quantify gene expression levels. Before the types of quantification methods are introduced, we will first take a closer look at the results generated.

Recall from Lab Session 3 that PCR amplifies DNA exponentially, with the amount of DNA doubling after each cycle. However, when reagents in the PCR sample become limiting, the amplification reaches a plateau. You cannot accurately quantify cDNA levels if you are analyzing in this plateau phase. Accurate quantification occurs during the early cycles of PCR, during exponential amplification. It is easy to obtain data from the exponential phase using qPCR because data was collected at every cycle, not just at the end.

In Bio-Rad's MyiQ software, each sample's amplification plot is shown in a different color. Fluorescence versus cycle number can be displayed in either the linear view or the log view (Figure 17.1). In either case, you can clearly see that fluorescence intensity (indicating DNA abundance) increases as cycle number increases. An important concept to keep in mind is that the more input cDNA that went into the PCR, the sooner an increase in fluorescence will occur. In Figure 17.1A (linear view), all samples produce minimal fluorescence during the first 13 cycles. These cycles are considered the baseline. Then, the blue amplification plot begins to increase significantly, followed by the red and green plots at later cycle numbers. Since the blue plot showed an increase in fluorescence at the earliest cycle number, we can conclude that the sample it represents contains more input cDNA (in this case, *23S*) than the other samples. To assign numeric values to these observations, a threshold is set, intersecting all the plots during exponential amplification. The threshold in Figure 17.1 is indicated by an orange horizontal line. Notice that in the linear view, each plot crosses the threshold in the early cycle numbers. Keeping all settings the same but viewing the data on a logarithmic scale (Figure 17.1B) we see that the threshold is high enough to be above background, but low enough to stay out of the plateau phase. The cycle at which a plot crosses the threshold is called the threshold cycle (C_T). The lower the C_T, the more abundant the template is. During exponential amplification, DNA should in theory double with each cycle. So, after 2 cycles, there will be 4 times as much as you started with, and after about 3.3 cycles, there will be 10 times as much. Even though the C_Ts of the amplicons shown in Figure 17.1B differ by only about three cycles, these represent 10-fold differences in *23S* amplicon abundance. If the same reactions were performed using conventional PCR, we may have analyzed amplicon abundance after 30 or more cycles on an agarose gel. Based on the qPCR results shown below (especially in the log view) it becomes clear that by this point in the reaction, the plateau phase has been reached and reagents have become limiting. We would conclude that there is very little difference in *23S* rRNA levels between samples. However, by examining

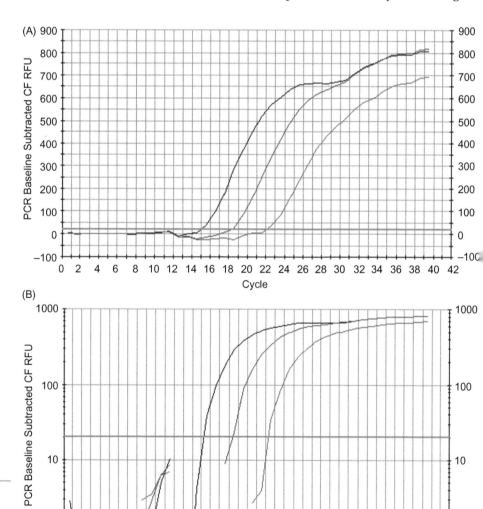

FIG. 17.1

Amplification plots. Ten-fold serial dilutions of cDNA were used as templates for RT-qPCR reactions using *23S* primers. The horizontal orange line represents the threshold. (A) Linear view. (B) Log view.

amplicons early in the reaction, in the exponential phase, we are able to determine that there are actually 10-fold differences between each sample.

There are two general methods for reporting quantification: absolute and relative. Absolute quantification calculates the quantity of a target gene in your unknown samples by interpolation from an absolute standard curve. This curve is commonly made by *in vitro* transcribed RNA or plasmid DNA with known copies of a target gene. This method is useful when you want to determine the exact copy number of a gene. The second type of quantification is called relative quantification. This aims to determine the difference in expression of a target gene between two samples. Reporting relative quantities is often sufficient when studying changes in gene expression. Basically, it aims to answer the question "How much of gene X is present in sample 1 compared to sample 2?" Relative quantification reports n-fold changes of gene expression rather than exact quantities: "There is 5-fold more of gene X in sample 1 versus sample 2." In this example, sample 2 serves as a calibrator (the sample used for the basis of a comparison). In our experiments, we will be comparing *gst::egfp* levels between no induction, induction with IPTG and induction with lactose. Therefore, samples with no induction will serve as calibrators, and *gst::egfp* levels of samples induced with IPTG or lactose will be expressed relative to those with no induction. Two methods exist for relative quantification: the relative standard curve method and the comparative C_T method. In the relative standard curve method, C_T values obtained from experimental samples are compared to C_T values from a standard curve made with dilutions of cDNA. The exact quantity of a target gene in the standards need not be known; just known mass amounts are required. The same procedure is also followed for a reference gene. Quantities are still expressed in comparison to a calibrator, unlike absolute quantification. The second method for relative quantification is the comparative C_T method (also called the $\Delta\Delta C_T$ method). This is the method that we will use in this lab session. It has an advantage in that no standard curve is required. Instead, mathematical formulas are used to calculate relative gene expression. It should be noted that the comparative C_T method assumes that PCR efficiencies are very similar between amplification of the target and reference genes. A validation experiment was performed previously to confirm that the primers used had very similar PCR efficiencies.

In our experiment, remember that we amplified not only *gst::egfp* (our target gene) but also *23S* rRNA (our reference gene). This reference gene is needed for the comparative C_T method. We will be comparing C_T values of *egfp* reactions to *23S* rRNA reactions for normalization purposes. Then, these differences (ΔC_T) will be compared to the calibrator, which is no induction. In this way, *gst::egfp* levels will be 1× in the no induction control and *gst::egfp* levels will have a n-fold difference with IPTG and lactose induction.

STEP 1: NORMALIZE TO AN ENDOGENOUS REFERENCE

For each cDNA in our experiment (no induction, IPTG and lactose), we will be taking the C_T of the *egfp* reaction minus the C_T of the *23S* reaction to give us ΔC_T.

$$\Delta C_T = C_T \text{ target gene} - C_T \text{ reference gene}$$

egfp   *23S*

STEP 2: COMPARE TO A CALIBRATOR

Once a ΔC_T value is calculated for no induction, IPTG and lactose cDNAs, we will report each value relative to the calibrator (no induction) to give us $\Delta\Delta C_T$. Notice that $\Delta\Delta C_T$ will always be zero for the calibrator sample itself.

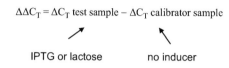

$$\Delta\Delta C_T = \Delta C_T \text{ test sample} - \Delta C_T \text{ calibrator sample}$$

IPTG or lactose   no inducer

STEP 3: CALCULATE RELATIVE QUANTITY

Finally, the amount of target, which is now normalized to an endogenous reference and relative to a calibrator, is calculated to provide us with the relative quantity of *gst::egfp*.

$$2^{-\Delta\Delta C_T}$$

With these values in hand, you can draw conclusions as to how much *gst::egfp* mRNA was expressed with IPTG and lactose in your positive clone. The references at the end of this chapter provide detailed guides on performing and analyzing RT-qPCR experiments.[1-4]

Laboratory Exercise

Relative Quantification of *gst::egfp* Levels

Note: The following instructions are specific for analyzing data obtained on the Bio-Rad iCycler using MyiQ Software. If you are using equipment made by another manufacturer, your instructor will provide you with data analysis instructions. Your instructor may perform steps 1–17 for you.

1. Open MyiQ software on the computer connected to the iCycler.
2. Click the "view post-run data" tab.
3. Double-click the data file (.odm) from your lab session to open it.
4. Click the "select wells" button and select the positions of your station's reactions. If your group had 12 reactions occupying the same row, you can click on the row's label (A, B, etc.) to automatically select that entire row. Click the "analyze selected wells" button.
5. Make sure that the "log view" button is displayed. That means that the current view is the normal view.
6. Verify that the amplification plots are running along 0 RFU (parallel to the x-axis) in the early cycles of the experiment. If the plots dip well below 0 or if RFU values vary greatly in the early cycles, you will need to adjust the baseline setting. To do this, determine the cycle at which your first (left-most) plot begins to rise above 0 RFU (amplification). Set the baseline cycles to between 2 and this cycle number.
7. Click the "log view" button.
8. Verify the auto-calculated threshold position (orange line) is high enough on the y-axis to be above any background noise (it should not intersect any sharp peaks seen in the early cycles) but is low enough to intersect the plots in the bottom third of the exponential phase.

Table 17.1 Pasting C$_T$ values into Excel

	A	B	C	D	E	F
1		*egfp*	*23S*			
2		C$_T$	C$_T$			
3	No inducer					
4	IPTG					
5	Lactose					

Table 17.2 Comparative C$_T$ method formulas

	A	B	C	D	E	F
1		*egfp*	*23S*			
2		C$_T$	C$_T$	ΔC$_T$	$\Delta\Delta$C$_T$	$2^{-\Delta\Delta C_T}$
3	No inducer			= B3 − C3	= D3 − D3	= 2^ − E3
4	IPTG			= B4 − C4	= D4 − D3	= 2^ − E4
5	Lactose			= B5 − C5	= D5 − D3	= 2^ − E5

9. If any adjustments are made in steps 6 and/or 8, you will need to click the "recalculate threshold cycles" button.
10. Click the "Reports" button.
11. In the "Select Report" drop-down menu, select "PCR baseline."
12. Click "save to file" and save the report to a thumb drive.
13. Click the "print" button to print a copy for your lab notebook.
14. Open Microsoft Word. Open the report you saved (a Rich Text Format file).
15. Copy the well number and C$_T$ values for each of your reactions.
16. Open Microsoft Excel.
17. Create a new spreadsheet and paste what you copied from the Word document. You may want to paste the values towards the bottom or right of the screen so that you can organize the rest of your spreadsheet (beginning with cell A1).
18. Create column and row headings as shown in Table 17.1. **Enter your C$_T$ values from +RT reactions only** in the appropriate cells by copying and, pasting.
19. Enter the formulas shown in Table 17.2 to calculate Δ C$_T$, $\Delta\Delta$ C$_T$ and $2^{-\Delta\Delta C_T}$ values.
20. Column F now shows the level of *gst::egfp* in each sample, relative to no inducer.
21. Below the data you have entered, paste in your C$_T$ values for the six −RT reactions.
22. For each RNA sample (no inducer, IPTG, lactose), locate the C$_T$ values corresponding to its −RT and +RT reactions for each primer pair. Remember that these −RT reactions were included to control for DNA contamination. Do the C$_T$ values differ between the −RT and +RT

reactions? Why? Remember that ideally DNA quantity doubles with each PCR cycle.

23. To calculate the amount of genomic DNA that is attributable to amplification in your +RT samples, perform the ΔC_T calculation for each *23S* reaction. This time, you will use the +RT sample as a calibrator:

$$\Delta C_T = C_{T\,without\,RT} - C_{T\,with\,RT}$$

Amplification with the *23S* primers, but not the *egfp* primers, will indicate genomic DNA contamination. Why? You are not interested in comparing the amount in one RNA sample relative to another, so you can skip the $\Delta\Delta C_T$ calculation. The equation $2^{-\Delta Ct}$ will calculate the quantity (%) of contaminating DNA. Multiply this value by 100 to obtain a percentage.

$$gDNA\;contamination_{IPTG}\;(\%) = 2^{-[(C_T\,without\,RT)-(C_T\,with\,RT)]} \times 100$$

24. To calculate the amount of plasmid DNA that is attributable to amplification in your +RT samples, perform the calculations described above for each *egfp* reaction.

References

1   Applied Biosystems. Guide to performing relative quantitation of gene expression using real-time quantitative PCR. <http://www3.appliedbiosystems.com/cms/groups/mcb_support/documents/generaldocuments/cms_042380.pdf/>, 2008.

2   Applied Biosystems. User bulletin #2. <http://www3.appliedbiosystems.com/cms/groups/mcb_support/documents/generaldocuments/cms_040980.pdf/>, 2001.

3   Applied Biosystems. Essentials of real time PCR. <http://www3.appliedbiosystems.com/cms/groups/mcb_marketing/documents/generaldocuments/cms_039996.pdf/>.

4   Schmittgen TD, Livak KJ. Analyzing real-time PCR data by the comparative C_T method. *Nat. Protoc.* 2008;3:1101–1108.

Discussion Questions

1. Why do standard PCR programs often include 30 cycles or more, while qPCR amplification plots (like the one in Figure 17.1B) show that at these later cycles, the plateau phase is often reached?

2. You were able to determine the levels of *gst::egfp* mRNA in the induced cultures relative to the non-induced. Without an inducing molecule to derepress the lac operator of pET-41a, was any *gst::egfp* mRNA transcribed? Why or why not?

LAB SESSION 18

Analysis of *gst::egfp* mRNA Levels by Semi-Quantitative RT-PCR: Part 1

Goal: Total RNA that you purified last week will be converted to cDNA by a reverse transcription reaction. That cDNA will then be used as a template for semi-quantitative PCR. You will amplify DNA with primers specific to *egfp* (your target gene) and *23S* rRNA (your reference gene).

Introduction

Read the Introduction in Lab Session 16 for information on reverse transcription. Semi-quantitative reverse transcription polymerase chain reaction (RT-PCR) is, as its name would suggest, less quantitative than RT-qPCR (see Lab Sessions 16 and 17), but can be used to detect changes in gene expression if they are approximately 10-fold or more. More subtle changes may be detected; however, the magnitude may be dampened due to the limited sensitivity of the assay. RT-PCR is considered an "end-point" method because you examine cDNA levels, and by inference mRNA levels, at the end of the PCR steps by agarose gel electrophoresis. You do not monitor cDNA levels throughout the PCR as you do in RT-qPCR. Recall from Lab Session 3 that PCR amplifies DNA exponentially, with the amount of DNA doubling after each cycle. However, when reagents in the PCR sample become limiting, the amplification reaches a plateau. You cannot accurately quantify cDNA levels if you are analyzing in this plateau phase. Accurate quantification occurs during the early cycles of PCR, during exponential amplification. Therefore, it is important to determine the exponential phase (the cycle numbers) when using RT-PCR to quantify gene expression. In practice, the exponential phase should be determined with each new cDNA sample with each primer pair before performing the actual experiment. This is most critical when trying to detect small-fold changes, but in our case we are examining large-fold changes. Furthermore, a range of cycles will generally fall within the exponential phase, so for the following experiment, it has been predetermined. There are several other reasons why semi-quantitative RT-PCR is considered semi-quantitative. First, detection of amplified cDNAs is based on the intensity of bands on an agarose gel. Ethidium bromide or fluorescent dyes like GelRed are reagents used to detect DNA; however, these are not very sensitive. Additionally, comparing intensities of different

PCR products is not very quantitative since a smaller product is typically amplified with higher efficiency than a larger product. To control for this limitation, the regions that you amplify for both genes (*egfp* and *23S* rRNA) will be the same length. Additionally, the amount of dye that is available to bind DNA will be kept constant by running all samples on the same gel. You should always avoid comparing band intensity from bands that are present on different gels. Finally, semi-quantitative RT-PCR relies on several successive steps being performed carefully in order to prevent imprecise results from sample to sample. This method, as well as RT-qPCR, assumes that the reverse transcription efficiency will be the same in each sample. In other words, if you pipette 1 µg of total RNA into each tube, the same amount of cDNA should be produced in each tube. Therefore, loading equivalent amounts of total RNA into each RT reaction is vital. To limit the possibility that reverse transcription efficiencies vary from sample to sample, a master mix of all reagents except for RNA will be aliquoted to ensure the equal treatment of each RNA. After this step, you must accurately pipette equivalent volumes of cDNA template into each PCR reaction. Again, we will utilize a master mix. Semi-quantitative RT-PCR includes a final step of agarose gel electrophoresis. You can imagine that if you underload a sample into the well, it will skew your quantification because this lane will have a lower intensity than if you had loaded it correctly. The *23S* products will serve as our loading control. Above all, sources of error can be minimized by careful pipetting. The major advantage of semi-quantitative RT-PCR over RT-qPCR is that is does not require a special thermal cycler capable of detecting fluorescence (which can be quite expensive).

Laboratory Exercises

Reverse Transcription (RT)

Please refer to instructions in Lab Session 16A.

Semi-Quantitative PCR

Note to instructor: A thermal cycler with 96 wells will hold PCR reactions for eight groups of students (12 reactions per group). If you have more than eight groups of students, you will need to either use a second thermal cycler or eliminate reactions in order for one thermal cycler to accommodate all the reactions. Preferably, eliminate the lactose-induced samples to leave 8 reactions instead of 12. The thermal cycler will then accommodate up to 12 groups. Since both *egfp* and *23S* need to be amplified for quantification and −RT reactions ensure contaminating DNA is not contributing to the quantification of mRNA, it is important to include these if at all possible.

The primers used to specifically amplify *gst::egfp* from your positive clone are:

- egfp 18 F CCTGAAGTTCATCTGCACCA
- egfp 18 R TGCTCAGGTAGTGGTTGTCG.

The primers used to specifically amplify *23S* rRNA from your positive clone are:

- 23S 18 F ACTGCGAATACCGGAGAATG
- 23S 18 R CCTGTTTCCCATCGACTACG.

The forward and reverse primers will be included together as a mix. Each primer pair should amplify a product 479 bp long.

1. The cDNA that you synthesized will now be used as a template in a semi-quantitative PCR assay. A very small volume of cDNA is needed (0.5 μl for a 25 μl reaction), so you will dilute each sample 1:5 in order to pipette a larger volume. Add 80 μl of water to each cDNA reaction tube (total volume is now 100 μl). Vortex to mix well. Repeat (with a fresh tip) for each cDNA reaction.
2. Obtain 12 PCR tubes and label them 1–12 on both the cap and side of the tube. Also label each tube with your station number. Place these in a PCR tube rack on ice.
3. Label two microcentrifuge tubes "*egfp* master mix" and "*23S* master mix." To ensure that multiple samples are treated equally with each reagent and for ease of pipetting, you will make a master mix consisting of each of the components listed below in Table 18.1. Each master mix will contain everything necessary for the PCR reaction except for the cDNA. One master mix will contain the primers specific for *gst::egfp*. The second master mix will contain the primers specific for *23S* rRNA.

Note to instructor: Table 18.1 shows the volumes needed for six PCR reactions with each primer pair (three +RT and three −RT reactions), including an additional 22.5 μl volume to account for pipetting error. If you decided to use a larger or smaller sample number, you will need to adjust volumes accordingly.

4. Pipette each reagent, in the order shown, into microcentrifuge tubes on ice.
5. Vortex well to mix and return to ice.
6. Pipette as accurately as possible 22.5 μl of the *egfp* master mix into each PCR tube labeled 1–6, changing pipette tips between each tube. Pipette 22.5 μl of the *23S* master mix into each PCR tube labeled 7–12, changing pipette tips between each tube.
7. Pipette as accurately as possible 2.5 μl of the appropriate diluted cDNA into the tubes as shown in Table 18.2. Again, change pipette tips between each tube.
8. Keep the tubes on ice until all stations are complete.
9. Place the tubes in a thermal cycler that has been programmed as shown in Table 18.3.

Table 18.1 PCR master mixes

Reagent	For 1 PCR reaction	*egfp* master mix	*23S* master mix
1. Sterile dH$_2$O	18.8 μl	131.6 μl	131.6 μl
2. 10× Taq buffer	2.5 μl	17.5 μl	17.5 μl
3. 10 mM dNTP mix	0.5 μl	3.5 μl	3.5 μl
4. Primer mix	0.5 μl	3.5 μl *egfp* primer mix	3.5 μl *23S* primer mix
5. Taq polymerase (5 U/μl)	0.2 μl	1.4 μl	1.4 μl
Total	22.5 μl	157.5 μl	157.5 μl

Table 18.2 PCR set-up

Tube	1	2	3	4	5	6	7	8	9	10	11	12
1. Master mix (22.5 µl)	*egfp*	*egfp*	*egfp*	*egfp*	*egfp*	*egfp*	*23S*	*23S*	*23S*	*23S*	*23S*	*23S*
2. cDNA (2.5 µl)	−no	+no	−IPTG	+IPTG	−lac	+lac	−no	+no	−IPTG	+IPTG	−lac	+lac

Table 18.3 PCR program

Cycle	Repeats	Step	Dwell time	Set point
1	1	1	2 minutes	95°C
2	26	1	30 seconds	95°C
		2	30 seconds	60°C
		3	45 seconds	72°C
3	1	1	10 minutes	72°C
4	1	1	∞	4°C

10. The program will take approximately 1.5 hours to complete. Your instructor will remove the tubes once the program is finished and store them at −20°C until your next lab session.

Discussion Questions

1. If you had primed reverse transcription of *E. coli* RNA with an oligo-dT primer rather than a random hexamer, how would that affect the experiment? Why?
2. You are amplifying both *gst::egfp* mRNA and *23S* rRNA in the PCR steps. These are two different types of RNA (messenger and ribosomal, respectively). Could an *E. coli* mRNA rather than an rRNA be amplified as a reference gene? Why? What is the main criterion for selecting a reference gene?

Analysis of *gst::egfp* mRNA Levels by Semi-Quantitative RT-PCR: Part 2

Goal: You will visualize semi-quantitative RT-PCR reactions by agarose gel electrophoresis. Imaging software (if available) will be used to quantify the levels of *gst::egfp* normalized to *23S* rRNA. In this way, you will be able to compare the relative levels of *gst::egfp* mRNA in *E. coli* induced with IPTG and lactose, compared to no inducer.

Introduction

Review Lab Session 18 Introduction.

Laboratory Exercises

Agarose Gel Electrophoresis

1. Pour a 1% agarose gel that includes GelRed or ethidium bromide. If 12 PCR reactions were performed, you will need two rows of wells (eight wells per row) on your minigel in order for all reactions and a molecular weight marker to fit. To do this, place one comb at the top of the gel casting tray as usual, and an additional comb in the middle of the gel casting tray. Alternatively, you could run all samples on a larger gel that will accommodate at least 13 wells across.
2. Obtain your PCR samples from the last lab session and allow them to thaw at room temperature.
3. Add 3 µl DNA loading buffer to each sample. Flick to mix.
4. Load 5 µl of each sample in the order shown in Table 19.1. Numbers indicate the label on the PCR tube (1–6 are *egfp* reactions, 7–12 are *23S* reactions).
5. Close the apparatus and connect the leads to the voltage pack. Run the gel at approximately 85 V or at an appropriate voltage for your apparatus (10 V/cm). When the top row's bromophenol blue dye-front has migrated one quarter of the way down the gel (half-way to the second row of wells), turn the power off and place the gel on a UV transilluminator. Photograph your gel and save it before proceeding.

Molecular Biology Techniques. DOI: 10.1016/B978-0-12-385544-2.00019-3

Table 19.1 Agarose gel loading order

Top row	1 kb ladder	1	2	3	4	5	6	empty
Bottom row	1 kb ladder	7	8	9	10	11	12	empty

Quantification

Note to instructor: if imaging software is not available, instruct students to visually compare band intensities of the *gst::egfp* PCR products (+RT) between the following samples: no inducer, IPTG and lactose. Also, have them observe and take into account any differences in band intensities of the *23S* rRNA PCR product. Presence and intensity of products in (−RT) lanes should also be recorded.

The following instructions refer to gels analyzed by Kodak Molecular Imaging Software.[1] Please refer to your software user's manual.

1. Open the Molecular Imaging Software.
2. Click "File", "Open" and select the Kodak MI Project (*.bip) file you want to analyze.
3. Click the "Image Display" button to invert your image and adjust intensity if necessary. Avoid analyzing saturated bands (indicated with red).
4. Click the "Lanes" button in the Navigation menu to the left.
5. Click the top button, "Set Search Area".
6. Select the area of interest. Resize the rectangular "Set Search Area" box so that it starts at the base of your top row of wells and encompasses all lanes. The bottom of the box should be below your bottom row of PCR products (*23S* rRNA).
7. Click the "New Lane Set" button on the left. Be sure that the lane lines run through the center of each band-containing lane. If a mini-gel was loaded as described in Table 19.1, each lane that the software finds will actually be two lanes on the gel, one directly above the other. Each lane the software finds should contain two bands: *egfp* in the top half of the gel and *23S* rRNA in the bottom half of the gel.
8. Click the "Find Bands" button on the left. Be sure that no bands show saturation (red). You may need to adjust the slider on the "Image Display" window to intensify/deintensify your image. Just click the "Image Display" button at the top if this window is not showing. You can also click the "Adjust Bands" button to change the sensitivity. Stop when all of your bands are labeled. You do not need to have any −RT bands labeled (if applicable). If there are extra "non-bands" labeled, just select them with the cursor and hit delete on the keyboard.
9. Click the analysis button at the top. You want to display the net intensity of each *gst::egfp* band and *23S* band in the +RT reactions only. Click "Display", "General", "Check intensities", "net intensity".
10. Net intensities should now be shown for each lane, with two numbers per lane. These numbers represent the net intensities of the *gst::egfp* band and the *23S* band. Print this table for your notebook.

11. Enter these net intensity values into Microsoft Excel. Calculate the relative *gst::egfp* mRNA levels in each sample with the following formulas:

   • Normalized *gst::egfp* level $= \dfrac{\text{Net intensity}_{egfp}}{\text{Net intensity}_{23S}}$

   • Relative *gst::egfp* level $= \dfrac{\text{Normalized } gst::egfp \text{ level}_{\text{with inducer}}}{\text{Normalized } gst::egfp \text{ level}_{\text{no inducer}}}$

12. Determine if there is any product in the −RT reactions. Amplification with the *23S* primers, but not the *egfp* primers, will indicate genomic DNA contamination. Why? Evaluate the level of DNA contamination by looking at each −RT reaction. Is some of the intensity you observed attributable to contaminating DNA?

13. Does the level of *gst::egfp* mRNA vary depending on the inducer? If so, was lactose or IPTG a more potent inducer?

Discussion Questions

1. You ran the agarose gel, but forgot to include the lactose-induced *23S* reaction containing reverse transcriptase (PCR tube 12). Without repeating the gel, what conclusions could you make about *gst::egfp* mRNA levels after lactose and IPTG induction? What would remain inconclusive? What if the sample you forgot to include was the lactose-induced *23S* reaction *without* reverse transcriptase (PCR tube 11)?

2. You were able to determine the levels of *gst::egfp* mRNA in the induced cultures relative to the non-induced. Without an inducing molecule to derepress the lac operator of pET-41a, was any *gst::egfp* mRNA transcribed? Why?

Reference

1 Kodak Molecular Imaging Software, Version 5.0 User's Guide. Carestream Health, Inc. 2008.

APPENDIX 1

Equipment

Shared Equipment

- −80°C degree freezer
- −20°C degree freezer (if the freezer has an automatic defroster, store all enzymes in insulated coolers, such as Stratacoolers, inside the freezer)
- refrigerator
- centrifuge
- beam balance (equal arm balance) for balancing centrifuge
- floor-model shaker incubator or table-top shaker incubators for 100 ml bacterial cultures
- 37°C incubator for bacterial plates and enzymatic reactions
- stir plate and stir bars
- plate spreader
- microwave oven for melting agarose
- sonicator with micro-immersion tip (such as the Fisher Sonic Dismembrator 50)
- Nanodrop spectrophotometer (*We very highly recommend the Nanodrop due to the minimal amount of sample that needs to be used; however, another spectrophotometer with quartz cuvette that measures absorbance at 260 and 280 nm could be used if DNA preps are scaled up appropriately.*)
- orbital bench-top shaker
- microtiter plate reader capable of reading EGFP fluorescence (excitation: 485 nm; emission: 528 nm)
- UV transilluminator with gel documentation system (or UV transilluminator in darkroom with camera)
- PCR thermocycler
- autoclave
- large water baths for cooling media to 60°C
- pH meter
- precision balance
- ice machine
- water purification system (such as Dracor or Millipore)
- real-time-capable thermal cycler
- each lab station should have the items detailed in the Station Checklist in Lab Session 1.

APPENDIX 2

Prep List

Notes to Prep Staff

Plasmids and *E. coli* Host Strains

The three plasmids you will need are pET-41a (Novagen cat. # 70556-3), pEGFP-N1 and pBIT (request from Dr. Sue Carson at Bit_Minor@ncsu.edu; include "pBIT or pEGFP-N1 request" in subject heading). The two host strains used are the expression host BL21(DE3) (Novagen cat. # 70235-3) and the non-expression host NovaBlue (Novagen cat. # 70181-3). The catalog numbers given are for chemically competent cells. If you choose to make competent cells, you may purchase regular, non-competent cells. Alternatively, other similar host strains may be purchased from other companies. Keep in mind that you will need one strain for isolating DNA (such as NovaBlue), and that strain should **not** be specialized for protein expression. You will need a second strain that allows transcription from the T7 promoter; this means that the expression strain must be an λ lysogen (such as BL21(DE3)).

Before the course begins, you will need a freezer stock of the following *E. coli* host strains (or similar strains) carrying the following plasmids.

E. coli host strain	Plasmid	Purpose
NovaBlue	pEGFP-N1	Template for PCR
NovaBlue	pET-41a	Plasmid isolation
BL21(DE3)	pET-41a	Negative control for fusion protein expression
BL21(DE3)	pBIT	Positive control for fusion protein expression

Before the course begins, you will need to purify maxiprep amounts of pEGFP-N1, pET-41a and pBIT.

Antibiotics

Each of the plasmids uses kanamycin for selection. Make a stock solution of kanamycin at 30 mg/ml in water, filter sterilize and store at −20°C. Add kanamycin to media at a dilution of 1:1000 from the stock solution (i.e. 1 ml in 1 liter) after autoclaving. It is critical to wait for media to cool to approximately 60°C before adding antibiotic. Adding antibiotic while the medium is too hot can cause inactivation of the antibiotic. It is also important to make sure that the antibiotic being used is effective; bad lots do occur. This can be tested by inoculating a freshly made LB/kan plate with wild type *E. coli* (lacking a plasmid containing antibiotic resistance).

Aliquots

Restriction enzymes and buffers should be aliquoted for individual use. Whenever feasible, make more aliquots than are needed for the class. Students often misplace aliquots or need to repeat a procedure. Extra aliquots come in handy. Additionally, extra volume is included in each aliquot to ensure students can pipette the full amount of reagent. Individual aliquots should be labeled either by hand or laser-printed labels.

Bacterial Waste

SOLID

Solid bacterial waste includes tubes, tips, plates, etc., that have been used to grow bacteria, or that have come in contact with any living bacteria/ biohazardous waste. Recombinant DNA should also be considered biohazardous. The prep staff should provide several containers with appropriate-sized BIOHAZARD bags (Fisher, Pittsburgh, PA). Replace the bags when a little over half full. All BIOHAZARD bags must be autoclaved before disposing. **Note**: Do not allow regular trash to be placed in BIOHAZARD bags. It can be dangerous to autoclave and puts an additional burden on the prep staff.

For convenience, students should have a small bench-top biohazard container, lined with an autoclave bag, at each station. This should be used for pipette tips and other small items that are used at the lab bench. Students should empty their small biohazard waste containers into the large common container at the end of each lab.

LIQUID

Prepare a carboy labeled "bacterial graveyard" for liquid bacterial waste only. Treat with bleach (approximately 15–20% by volume). Wait 10–20 minutes. Bacteria are dead if the liquid has cleared and cell debris has fallen to the bottom of the container. Decontaminated liquid may be poured down the drain.

Autoclaving

Autoclaved materials should be ready the day before the material is needed.

LIQUID

Fill bottles to only 60% full or less. Caps should be loose to prevent pressure build-up and bursting. Liquids should be sterilized for 20 minutes. Tighten the caps fully only after the solution has cooled to room temperature.

BIOHAZARD WASTE

Biohazard bags must be sealed and labeled. If leakage is a problem, they may have to be double bagged or absorbent material added (e.g., vermiculite, bentonite, etc.)

DRY GOODS

All pipette tips and microfuge tubes should be autoclaved on gravity/dry cycle before use.

General Lab Preparation

Designate an area or cart for reagents for student use. Group together materials needed for a particular lab session. Use 5 × 7 inch note cards to label all reagents and include information such as "for Lab 7A, step 5." Include phrases like "Return when finished," "Do not throw away, label with your group number and store at 4°C," or "Discard following use." Be sure to also label each individual aliquot, as students will often mix up tubes! Providing clear directions whenever possible will help avoid problems and is especially important to ensure a safe working environment.

Periodically check the BIOHAZARD bags and remove them for autoclaving when they are half full. Make sure that 1× TBE carboys and 70% ethanol carboys are full.

Students

Students will ask for help and for more reagents, particularly during the first several sessions. Advanced preparation and having additional aliquots on hand will help to avoid frustration.

Supplies and Reagents

Supplies and reagents for each week are listed in this appendix according to lab session. Volumes listed for reagents generally contain an extra ~10% to account for pipetting discrepancies. There are certain supplies and reagents that should be available at all times. These will not necessarily be listed under each lab session, but should be available for students and prep staff for every lab session. These are listed below:

Supplies and Reagents for General Use

- non-latex gloves (small, medium, large and extra-large)
- powder-free gloves are preferable for working with nitrocellulose membranes
- sterile pipette tips (all sizes)
- microcentrifuge tubes
- sterile distilled/deionized water
- 95% ethanol
- 70% ethanol (carboy)
- agarose (can be kept at a general use area containing top-loading balance and microwave)
- Petri dishes (square and round)
- 50 ml conical tubes
- 15 ml conical tubes
- weigh boats
- centrifuge tubes capable of being centrifuged at 15,000 × g
- sterile toothpicks for inoculating cultures
- LB broth

- LB/kan agar plates (stored at 4°C)
- 1× TBE (carboy)
- GelRed (10,000× ; should be stored in foil and protected from light)
- hand soap
- paper towels.

Recipes for General Use

LB BROTH AND LB AGAR

1. Mix the following:

Compound	Amount per liter
Tryptone	10 g
NaCl	10 g
Yeast extract	5 g
Deionized water	1000 ml

Alternatively, a premixed powder (such as Fisher Scientific catalog number BP1426-500) can be purchased and mixed directly with deionized water at 25 g per liter.

2. Autoclave.
3. Cool to 60°C and add filter-sterilized kanamycin (final concentration 30 µg/ml; can use 1000× stock solution) and/or filter-sterilized IPTG (final concentration 20 µg/ml; can use 1000× stock solution), where necessary.

To prepare **LB agar**, prepare LB broth then add 15 g/l agar and a stir bar before autoclaving (or use premixed LB agar, such as Fisher Scientific catalog number 1425–500). Autoclave, then place in a 60°C water bath. When the temperature has equilibrated, place on stir plate and add kanamycin and/or IPTG where necessary. Pour approximately 20 ml per Petri dish.

IPTG (20 mg/ml)

Dissolve 1.0 g IPTG in 50 ml deionized water. Filter sterilize. This is a 1000× stock. Store at −20°C.

KANAMYCIN (30 mg/ml)

Dissolve 1.5 g kanamycin in 50 ml deionized water. Filter sterilize. This is a 1000× stock. Store at −20°C.

5× TBE STOCK (TRIS-BORATE-EDTA)

Reagent	Amount
Tris	54 g
Boric acid	27.5 g
EDTA	4.15 g (or 20 ml 0.5 M EDTA, pH 8)
dH₂O	To 1 liter

Place in carboy for students to prepare 1× stocks for their own use. Prepare 6 liters for 20 groups.

Alternatively, 10× TBE can be ordered (for example, Genesee Scientific # 20-196) and diluted for the students prior to the start of the lab.

GelRed STOCK (10,000×)

Per group, 30 µl in tinted microcentrifuge tubes, or wrapped with aluminum foil since GelRed is light-sensitive. Students should save the GelRed at their lab station to use in multiple laboratory exercises.

DNA PRIMERS

DNA primers can be ordered from many commercial sources (e.g., Integrated DNA Technologies). They should be ordered desalted, but need no additional modifications. Upon receipt, prepare a stock solution of the primer by dissolving it in sterile water to a final concentration of 100 pmol/µl (100 µM).

Lab session	Primer name	Primer sequence
3	egfpNco	AAACCATGGTGAGCAAGGGCGA
3	egfpNot	AAAGCGGCCGCTTTACTTGTACA
7B	pBITfor	CAAGCTACCTGAAATGCTGA
7B	pBITrev	CTTGTACAGCTCGTCCATGC
9C	Sequencing primer	CGAACGCCAGCACATGGACAGC
16	egfp16 F	CGACGGCAACTACAAGACC
16	egfp16 R	GTCCTCCTTGAAGTCGATGC
16	23S16 F	GACGGAGAAGGCTATGTTGG
16	23S16 R	GTTGCTTCAGCACCGTAGTG
18	egfp 18 F	CCTGAAGTTCATCTGCACCA
18	egfp 18 R	TGCTCAGGTAGTGGTTGTCG
18	23S 18 F	ACTGCGAATACCGGAGAATG
18	23S 18 R	CCTGTTTCCCATCGACTACG

LAB SESSION 1

BSA Serial Dilutions and Nitrocellulose Spot Test

Note: This exercise will be done by each student, not by each group.

The students will each use the following supplies:

- sterile distilled/deionized water (at least 1 ml per student)
- 15 µl bovine serum albumin, 1 mg/ml, store at −20°C
- nitrocellulose membrane approximately 2″ × 3″ (1 sheet per group)
- Whatman 3MM paper, approximately 2″ × 3″ (1 sheet per group)
- square Petri dish, tip box top, or other convenient tray for staining (1 per group).

The following is to be shared:

- waste containers with funnels for used destaining solution
- Amido Black Stain
- destaining solution.

Recipes

Amido Black Stain: The following recipe provides one bottle of stain for general use. Stain may be reused if stored in an airtight container.

1. Combine the following:

	Amount	Final concentration
Naphthol blue black 10B (Sigma-Aldrich)	0.5 g	0.1%
Methanol	225 ml	45%
Acetic acid	35 ml	7%
Deionized water	To 1 liter	

2. Filter through fluted filter paper.
3. Store at room temperature.

Destaining solution: The following provides two bottles for general use; mix and store at room temperature.

	Amount	Final concentration
Methanol	700 ml	70%
Acetic acid	70 ml	7%
Deionized water	To 1 liter	

LAB SESSION 2

Purification and Digestion of Plasmid (Vector) DNA

Note to prep staff:

1. Two days before the lab, streak LB/kan plates with the cloning strain of *E. coli* (such as NovaBlue) containing the pET-41a plasmid.
2. The morning of the day before the lab, pick a single colony from the LB/kan plate and inoculate a starter culture of 2–5 ml LB medium containing kanamycin. Incubate for ~8 hours at 37°C with vigorous shaking (~300 rpm).
3. The afternoon of the day before the lab, dilute the starter culture 1:500 into 100 ml selective LB/kan medium and shake at 37°C in a 500 ml flask.

Have available for pre-lab set-up:

- LB broth containing 30 μg/ml kanamycin
- LB agar containing 30 μg/ml kanamycin
- *E. coli* strain NovaBlue (or other K12 strain) containing pET-41a.

The students will use the following supplies and equipment:

- QIAGEN® QIAprep Spin Miniprep Kit (catalog number 27104).

The following supplies and reagents are included in the QIAGEN® QIAprep Spin Miniprep Kit. Each kit contains columns and reagents for 50 plasmid preps. Be sure to follow the kit's protocol in adding the provided RNase to Buffer P1 and 100% ethanol to Buffer PE.

Aliquot the following per group:

- 275 μl Buffer P1, store at 4°C
- 275 μl Buffer P2

- 385 µl Buffer N3
- 800 µl Buffer PE
- 55 µl Buffer EB
- QIAprep spin column.

Aliquot the following per group for the restriction digest:

- 5.5 µl 10× restriction enzyme buffer (NEB buffer 3), store at −20°C
- 1.0 µl BSA (10 mg/ml), store at −20°C
- 2.5 µl *Nco*I, store at −20°C
- 2.5 µl *Not*I, store at −20°C
- 55 µl sterile dH$_2$O.

Students will purify their own pET-41a, but have extra DNA on hand just in case.

LAB SESSION 3

PCR Amplification of *egfp* from pEGFP-N1; Clean-up and Visualization of Digested pET-41a Vector

The students will use the following supplies. The following is needed per group:

- 50 µl sterile dH$_2$O
- 5.5 µl 10× Thermopol reaction buffer
- 100 ng pEGFP-N1 plasmid DNA (i.e. 2 µl of 50 ng/µl DNA)
- 5.5 µl dNTP mix (stock of 2 mM each dNTP), store at −20°C
- 1.2 µl egfpNco primer (AAACCATGGTGAGCAAGGGCGA; 100 pmol/µl), store at −20°C
- 1.2 µl egfpNot primer (AAAGCGGCCGCTTTACTTGTACA; 100pmol/µl), store at −20°C
- 0.55 µl Vent polymerase, store at −20°C
- PCR tube with dome-capped lid
- 30 µl GelRed stock (10,000×) (see recipe in "Supplies and Reagents for General Use"), students should save for future labs
- 500 µl 10× DNA loading buffer (students should save for future labs)
- 100 µl NEB 1 kb DNA ladder (NEB catalog number N3232), mixed with loading dye. Students should save for future labs.

The following supplies and reagents are included in the QIAGEN® QIAquick PCR Purification Kit (catalog number 28104). Each kit contains columns and reagents for 50 DNA "clean-ups."

Aliquot the following amounts for student use:

- 275 µl Buffer PB
- 800 µl Buffer PE
- 55 µl Buffer EB
- 1 QIAquick column and 2 ml collection tube.

Students will share the following items:

- agarose
- 1× TBE buffer (see recipe in "Supplies and Reagents for General Use").

Recipes

10× DNA LOADING BUFFER

Reagent	Amount needed for 50 ml
25% glycerol	12.5 ml
0.1 M EDTA	10 ml of 0.5 M EDTA
0.25% bromophenol blue	0.125 g

Aliquot 500 μl per tube, one tube per group. Students should save loading buffer for future labs.

NEB 1 kb DNA LADDER (NEB CATALOG NUMBER N3232)

Dilute one part DNA ladder with one part 10× DNA loading buffer, and eight parts sterile, deionized water. Aliquot 100 μl per group. Students should save extra 1 kb ladder in their freezer boxes for future labs.

If an alternative molecular weight ladder is used, follow manufacturer's instructions and supply students with a photograph of the ladder with molecular weights labeled.

LAB SESSION 4

Preparation of Insert DNA (*egfp*) PCR Product

The students will use the following supplies.
The following supplies and reagents are included in the QIAGEN®
QIAquick PCR Purification Kit (catalog number 28104). Aliquot per group:

- 550 μl Buffer PB
- 1.75 ml Buffer PE
- 110 μl Buffer EB
- two QIAquick columns and two 2 ml collection tubes.

Aliquot the following for the restriction digest:

- 55 μl sterile dH$_2$O
- 5.5 μl 10× restriction enzyme buffer (NEB buffer 3), store at −20°C
- 1.0 μl BSA (10 mg/ml), store at −20°C
- 1.2 μl *Nco*I, store at −20°C
- 1.2 μl *Not*I, store at −20°C.

Students will share the following reagents:

- agarose
- 1× TBE buffer
- sterile dH$_2$O for Nanodrop
- GelRed stock (10,000×).

LAB SESSION 5

DNA Ligation and Transformation of *Escherichia coli*

The students will use the following supplies. Aliquot the following per group:

- 2.5 μl 10× ligase buffer, store at −20°C
- 1.5 μl T4 DNA ligase, store at −20°C

Note: This protocol uses NEB (New England Biolabs) T4 DNA ligase, catalog number M0202L. If using a different brand of T4 DNA ligase, follow the manufacturer's instructions. The incubation time may be significantly longer.

- 75 µl competent *E. coli* BL21(DE3) cells

Note: Thaw frozen competent cells *immediately* before use. Mix gently before aliquoting. Cells should not sit on ice for more than a few minutes. Cells should not sit at room temperature for any period of time.

- 250 µl LB or SOC broth
- three LB/kan plates.

Students will share the following reagents:

- 95% ethanol for plate spreading
- plate spreader
- sterile dH$_2$O.

Students should have from previous labs (but have extra on hand, just in case):

- undigested pET-41a
- 1 kb DNA ladder
- 10× DNA loading buffer
- GelRed stock (10,000×).

LAB SESSION 6

Interim Lab

Note to prep staff: The afternoon of the day before meeting, streak out the following cultures on several LB/kan plates each:

- pBIT in host strain BL21(DE3)(positive control)
- pET-41a in host strain BL21(DE3)(negative control).

Streak in quadrants in order to ensure isolated colonies. The students will each need a colony of pBIT and pET-41a for their positive and negative controls, respectively. Students who did not obtain any transformants on their plates can use extra transformants from their classmates' plates.

The students will use the following supplies. The following is needed per group:

- two LB/kan plates
- two grid stickers (Diversified Biotech PetriSticker™ 32-square grid, catalog number PSTK-1000)
- sterile toothpicks
- Sharpies (or other marker pens)
- biohazard waste container (for toothpicks used to pick colonies)
- students' transformation plates from Lab Session 5.

Students will share the following items:

- several LB/kan plates containing pBIT in host strain BL21(DE3) (positive control)
- several LB/kan plates containing pET-41a in host strain BL21(DE3) (negative control)
- labeling tape (to tape their plates together).

LAB SESSION 6

Colony Hybridization: Monoclonal Antibody Probe

Note to prep staff:

1. Place students' replica plates inverted in the refrigerator (lids down) until the afternoon prior to the lab session.
2. The afternoon prior to the lab session, place students' replica plates (lids down) in the 37°C incubator.

The students will use the following supplies. The following is needed per group:

- one circle of nitrocellulose
- 750 µl IPTG solution (20 mg/ml in H_2O, filter sterilize)
- LB/kan plate
- one piece of plastic wrap
- glass Petri plates with support mesh (or toothpicks taped to the edges to hold plastic Petri dishes above)
- 8 ml lysis buffer, store at 4°C
- 10 ml blocking solution, store at 4°C
- 10 ml 1× wash buffer/IGEPAL/milk, store at 4°C
- 10 ml 1× wash buffer/milk, store at 4°C
- three squares Whatman paper
- empty Petri dishes
- pencil (for writing on nitrocellulose).

Students will share the following items:

- chloroform, with a glass pipette for pipetting into Petri dishes.

Recipes

BLOCKING SOLUTION

Combine the following. Aliquot 10 ml into 15 ml conical tubes, one per group, and freeze at −20°C. This can be thawed at 4°C the week of the lab. Note: Extra blocking solution will be needed to make Lysis buffer.

Component	Amount added	Final concentration
Nonfat powdered milk*	15 g	3%
NaCl, 1 M	18.75 ml	75 mM
Tris (pH 8.0), 1 M	6.25 ml	25 mM
Deionized water	To 500 ml	

*Nonfat powdered milk is available at most grocery stores. Bovine serum albumin (BSA; fraction V) can be substituted for nonfat powdered milk.

LYSIS BUFFER

Aliquot 8 ml into 15 ml conical tubes, one per group. This can be stored at 4°C for the week of the lab.

Blocking solution	250 ml
Lysozyme	10 mg
DNase (1 mg/ml)*	25 µl
$MgCl_2$, 1 M	1.25 ml

*Pancreatic DNase (1 mg/ml), prepare in deionized, sterile water and store at −20°C. Use to make lysis buffer.

1× WASH BUFFER/MILK AND 1× WASH BUFFER/IGEPAL*/MILK

For both recipes, first make a 10× stock:

10× WASH BUFFER

Tris (pH 7.4), 1 M	200 ml
NaCl	87.6 g
Deionized water	To 1 liter

1× WASH BUFFER/MILK

Aliquot 10 ml per group.

Component	Amount added	Final concentration
10× wash buffer	100 ml	1×
Deionized water	900 ml	
Nonfat powdered milk	1 g	0.1%

1× WASH BUFFER/IGEPAL/MILK

Add 0.5 ml IGEPAL to 1 L of "1× wash buffer/milk" (this is a 0.05% final concentration of IGEPAL). Aliquot 10 ml per group.

LAB SESSION 7

Characterization of Recombinant Clones

The students will use the following supplies. The following is needed per group:

- 10 ml 1× wash buffer/IGEPAL/milk (same recipe as in Lab Session 6), store at 4°C
- 70 ml 1× wash buffer/milk (same recipe as in Lab Session 6), store at 4°C
- 8.5 μl α-GFP antibody (Clontech catalog number 632381), store at −20°C, but limit freeze–thaw cycles
- 15 ml blocking solution (same recipe as in Lab Session 6), store at 4°C
- 16.5 μl goat anti-mouse conjugated to peroxidase (GAMP) (Sigma-Aldrich catalog number A3673), store at −20°C, but limit freeze–thaw cycles
- 7.5 ml peroxide stain (from chloronaphthol stock), make fresh the day of the lab
- one empty Petri dish
- paper towels
- one LB/kan plate
- one set of PCR strip tubes and caps
- PCR master mix (180 μl per group).

---

*IGEPAL is a nonionic, non-denaturing detergent that is also known as IGEPAL CA-630.

Item	Per group (180 μl)
Sterile dH₂O	155.2 μl
dATP (100 mM stock)	0.36 μl
dTTP (100 mM stock)	0.36 μl
dCTP (100 mM stock)	0.36 μl
dGTP (100 mM stock)	0.36 μl
pBITrev 100 pmol/μl*	0.9 μl
pBITfor 100 pmol/μl*	0.9 μl
10× buffer with Mg	18 μl
Taq polymerase	3.6 μl

Note: If your PCR buffer does not contain Mg, add it according to the manufacturer's instructions.

- pBITrev: CTTGTACAGCTCGTCCATGC
- pBITfor: CAAGCTACCTGAAATGCTGA.

Recipes

CHLORONAPHTHOL STOCK SOLUTION

This is used to make the peroxide stain. Wear gloves – chloronaphthol is a suspected carcinogen. Combine the following and store at −20°C in a bottle protected from light:

4-Chloro-1-naphthol	150 mg	(0.3% final concentration)
Ice-cold methanol	50 ml	

PEROXIDE STAIN

This should be made fresh immediately before use. Purchase new hydrogen peroxide each semester.

Component	Amount added	Final concentration
Chloronaphthol stock	17 ml	0.05%
Deionized water	68 ml	
Tris (pH 7.4), 1 M	5 ml	50 mM
NaCl, 1 M	10 ml	100 mM
H₂O₂,* 30%	35 μl	0.02%

*Hydrogen peroxide (H₂O₂) comes as a 30% solution and is converted to H₂O so it should be bought fresh each semester.

LAB SESSION 8

Interim Lab

The students will use the following supplies. The following are needed per group:

- six 2 ml aliquots of LB/kan broth (in snap cap culture tubes)
- sterile toothpicks
- tube racks
- Sharpies (or other marker pens)
- biohazard waste container
- access to their replica plate (replicated master plate on LB/kan from Lab Session 7).

LAB SESSION 8

Characterization of Recombinant Clones: Part 2

The students will use the following supplies.

The following supplies and reagents are included in the QIAGEN® QIAprep Spin Miniprep Kit (catalog number 27104). Each kit contains columns and reagents for 50 plasmid preps. Be sure to follow the kit's protocol in adding the provided RNase to Buffer P1 and 100% ethanol to Buffer PE.

The following is needed per group:

- 1.75 ml Buffer P1, store at 4°C
- 1.75 ml Buffer P2
- 2.45 ml Buffer N3
- 5.25 ml Buffer PE
- 350 µl Buffer EB
- six QIAprep spin columns
- one LB/kan/IPTG plate
- one LB/kan plate
- sterile toothpicks.

Students will share the following:

- agarose
- $1\times$ TBE buffer.

Students should have from before (but have extra on hand, just in case):

- 1 kb DNA ladder
- $10\times$ DNA loading buffer
- GelRed stock ($10,000\times$).

LAB SESSION 9

Characterization of Recombinant Clones: Part 3

The students will use the following supplies. Aliquot the following per group:

- 16 µl $10\times$ restriction enzyme buffer 3, store at $-20°C$
- 1.8 µl BSA (10 mg/ml), store at $-20°C$
- 8 µl *Nco*I, store at $-20°C$
- 8 µl *Not* I, store at $-20°C$
- 1.2 µl uncut pBIT DNA (~100 ng/µl)
- 3 µl of 10 pmol/µl DNA sequencing primer (CGAACGCCAGCACATGGA CAGC). Dilute stock solution 1:10 in sterile water to obtain a working solution of 10 pmol/µl (10 µM).
- PCR tube
- sterile dH$_2$O.

Students will share the following:

- agarose
- 1× TBE.

Students should have from before (but have extra on hand, just in case):

- 1 kb DNA ladder
- 10× DNA loading buffer
- GelRed stock (10,000×)
- students' miniprep DNA from Lab Session 8
- IPTG replica plate from Lab Session 8.

LAB SESSION 10

Computational Analysis of DNA Sequence from a Positive Clone: Part 2

Bioinformatics exercise where students need data from their sequencing reaction from Lab Session 9 and a computer with Chromas Lite software for analysis. http://www.technelysium.com.au/chromas_lite.html.

LAB SESSION 11

Interim Lab

The students will use the following supplies:

- three snap-cap tubes with 1 ml LB/kan
- racks for tubes
- sterile toothpicks
- Sharpie markers
- master replica plates (the most freshly streaked LB/kan plate).

LAB SESSION 11

Expression of Fusion Protein from Positive Clones, SDS-PAGE and Western Blot: Part 1

Note to prep staff:

1. Two days before the laboratory: streak LB/kan plates of the positive and negative control strains.
   - positive: pBIT in *E. coli* strain BL21(DE3) or other appropriate expression strain
   - negative: pET-41a in *E. coli* strain BL21(DE3) or other appropriate expression strain.
2. One day before the lab, incubate the students' LB/kan broth cultures overnight at 37°C. Also inoculate and incubate 10 ml each of the positive and negative controls into LB/kan broth. The positive and negative controls are the pBIT and pET-41a plasmids respectively, in *E. coli* host strain BL21(DE3). Inoculating 10 ml of each control will result in enough bacteria for 30 lab stations.

3. Three to four hours before the start of the lab, add 2 ml 2× YT/kan/IPTG to each 1 ml student culture and continue to shake at 37°C. Also, add 2× YT/kan/IPTG in the same ratio to each of the controls (i.e. 20 ml 2× YT/kan/IPTG to 10 ml culture) and continue shaking.

Prep staff use:

- 2× YT broth with kanamycin and IPTG.

The students will use the following supplies. The following is needed per group:

- student cultures that have been incubated in 2× YT/kan/IPTG 3–4 hours before the lab
- five locking or screw-cap microfuge tubes, or caps that fit over regular microfuge tubes that prevent them from popping open during boiling
- one polyacrylamide gel (use a gel with a polyacrylamide concentration in the range of 10–12%, such as Bio-Rad catalog numbers 161-1101 or 161-1102, or ISC BioExpress catalog numbers E-4325-010 or E-4325-012)
- 1 ml 2× sample (loading) buffer (students should save extra)
- 10 μl MW marker (NEB prestained broad range molecular weight marker, catalog number P7708S)
- two pieces Whatman 3MM paper, cut to the size of the fiber pads
- one piece nitrocellulose, cut to the size of the gels
- two fiber pads (NOT disposable)
- one glass casserole dish.

The following is to be shared:

- pBIT and pET-41a *E. coli* BL21(DE3) cultures that have been incubated in 2× YT/kan/IPTG 3–4 hours before the lab
- 1× Tris-glycine running buffer
- sodium bicarbonate
- one pack of disposable square Petri dishes (Fisher 08-757-11A)
- one pack of glass tubes or disposable pipettes for use as "rolling pins."

Recipes

2× YT BROTH

Component	Amount needed for 20 groups
Bacto tryptone	16 g
Yeast extract	10 g
NaCl	5 g
Deionized water	To 1 liter

Autoclave and store at room temperature.

2× YT/KAN/IPTG

To 100 ml 2× YT, add 100 μl kanamycin (30 mg/ml stock) and 100 μl IPTG (20 mg/ml stock). The final concentration of kanamycin is 30 μg/ml, and the final concentration of IPTG is and 20 μg/ml.

173

TRIS–GLYCINE RUNNING BUFFER

Recipes for the 1× concentration and a 10× stock solution are below:

	Amount		Final concentration	
	1×	**10×**	**1×**	**10×**
Tris base	6.0 g	30 g	25 mM	0.25 M
Glycine	28.8 g	141 g	190 mM	1.9 M
SDS	2.0 g	10 g	0.1%	1%
Deionized water	To 2000 ml	To 1000 ml		

Place carboys of 1× running buffer in the lab. It is a good idea to keep extra 10× stock on hand.

2× SAMPLE (LOADING) BUFFER

Prepare under the fume hood:

	Amount	Final concentration
SDS	1 g	1%
Tris (pH 6.8), 0.5 M	25 ml	125 mM
Sucrose	15 g	15%
β-mercaptoethanol	10 ml	10%
EDTA (pH 7.0) 0.1 M	1 ml	1 mM
Bromophenol blue, 1%	5 ml	0.05%
Deionized water	To 100 ml	

Aliquot into 1 ml volumes (one per group) and store frozen.

LAB SESSION 12

Expression of Fusion Protein from Positive Clones, SDS-PAGE and Western Blot: Part 2

The students will use the following supplies. The following is needed per group:

- 10 ml blocking solution (TBS-T plus 5% powdered milk), store at 4°C
- 8.5 μl α-GFP antibody, store at −20°C, but limit freeze–thaw cycles
- 16 μl goat anti-mouse antibody conjugated to peroxidase (GAMP) (Sigma-Aldrich catalog number A3673), store at 4°C
- 10 ml peroxide stain (made fresh from chloronaphthol stock, same recipe as in Lab Session 7) (*30% hydrogen peroxide should be less than 1 month old.*)
- two pieces of Whatman 3MM paper, cut to the size of the fiber pads
- one LB/kan plate
- sterile toothpicks.

The following is to be shared:

- Ponceau S stain (Sigma-Aldrich catalog number P7170), approximately 10 ml needed per group (*Ponceau S stain may be saved and reused*)
- one pack of disposable square Petri dishes (Fisher 08-757-11A)
- TBS-T, at least 150 ml needed per group for washes.

Recipes

TBS (TRIS-BUFFERED SALINE, PH 7.6)

	Amount	Final concentration
Tris base	2.42 g	20 mM
NaCl	8 g	137 mM
HCl (1 M), to pH 7.6	3.8 ml	10%
Deionized water	To 1 liter	

TBS-T: Add Tween 20 to TBS to a final concentration of 0.1%. Prepare a large amount for common use.

Blocking solution: TBS-T plus 5% nonfat milk powder (or bovine serum albumin fraction V).

LAB SESSION 13

Interim Lab

The students will use the following supplies. The following is needed per group:

- sterile toothpicks
- one LB/kan 2 ml aliquot in snap cap culture tube
- sharpie marker
- students' replica plates of positive clone (streaked last lab session)
- rack for culture tubes.

LAB SESSION 13

Extraction of Recombinant Protein from *Escherichia coli* Using a Glutathione Affinity Column

Note to prep staff:

1. The afternoon of the day before the lab: incubate students' cultures (that were stored in the refrigerator after the interim lab) overnight shaking at 37°C.

    Also inoculate and incubate E. coli *strain BL21(DE3) harboring the pBIT plasmid. This can be used as a backup for purification if students accidentally spill their own cultures.*
2. Three to four hours prior to the start of the lab: subculture students' 2 ml cultures and backup culture into 250 ml flasks containing 100 ml 2 × YT/kan/IPTG each. Incubate shaking at 37°C.

The students will use the following supplies. The following is needed per group:

- two 50 ml centrifuge tubes capable of withstanding 10,000 × g force
- 100 ml 2× YT/kan/IPTG (in 250 ml flasks for prep staff to use for subculturing)
- 20 µl lysozyme (10 mg/ml), store at −20°C
- GST-Bind resin: 0.75 ml settled bead volume in 15 ml conical tube (Novagen catalog number 70541-4, or equivalent GST affinity resin), store at 4°C
- one chromatography column and stopcock (Bio-Rad catalog numbers 731-1550 and 732-8102)

- 20 ml GST Bind/Wash buffer, room temp (a component of the Novagen GST-Bind Buffer kit, catalog number 70534-3, or equivalent) (*Dilute 10× stock according to the manufacturer's instructions*)
- 800 μl GST Bind/Wash buffer with Pefabloc, ice-cold (*Pefabloc® 4-(2-Aminoethyl) benzenesulfonyl fluoride hydrochloride, AEBSF, Sigma-Aldrich catalog number 76307), 0.1 M stock solution (prepare stock solution with sterile, deionized water, and freeze at −20°C*)
- 3.5 ml GST elution buffer (a component of the Novagen GST-Bind Buffer kit, catalog number 70534-3, or equivalent) (*Prepare from 10× Glutathione Reconstitution Buffer and reduced glutathione, according to manufacturer's instructions*).

The following is to be shared:

- LB for resuspending bacteria (a few small bottles per lab or one 1.3 ml tube per group)
- dry ice
- a refrigerated microfuge or place several microfuges in a refrigerator at 4°C.

Have on hand in case students run out or accidentally threw away their extra:

- 2× SDS-PAGE sample buffer.

Recipes

GST BIND/WASH BUFFER WITH PEFABLOC®*

Add 1 ml of the 0.1 M Pefabloc® stock solution to 50 ml 1× Bind/Wash buffer. Mix and keep at 4°C until used.

LAB SESSION 14

Analysis of Purification Fractions

The students will use the following supplies. The following is needed per group:

- 7 ml PBS (phosphate buffered saline), pH 7.4
- 3.2 μl of recombinant EGFP: 1 mg/ml solution in PBS (Biovision, Inc. Catalog number 4999-100)
- 96-well plate
- one polyacrylamide gel (use a gel with a polyacrylamide concentration in the range of 10–12%, such as Bio-Rad catalog numbers 161-1101 or 161-1102, or ISC BioExpress catalog numbers E-4325-010 or E-4325-012)
- 10 μl prestained molecular weight marker for SDS-PAGE
- 20 ml Gelcode Blue staining reagent (Pierce catalog number 24590), store at 4°C; order fresh every semester
- square Petri dish
- locking-cap microfuge tubes or cap locks to prevent microfuge tube lids from popping open during heating (Fisher catalog number NC9346739)
- one LB/kan plate.

*Pefabloc® is a protease inhibitor. Wear gloves, eye protection and lab coat when working with this chemical.

The following will be shared:

- 1× Tris-glycine running buffer, recipe as described in Lab Session 11.

Have on hand in case students run out:

- 2× SDS-PAGE sample buffer.

Students should have from the previous week:

- cell lysate and fractions collected from purification.

Recipes

1× PHOSPHATE BUFFERED SALINE (1× PBS), PH 7.4

	Amount	Final concentration
KCl	0.2 g	2.7 mM
NaCl	8 g	137 mM
Na$_2$HPO$_4$	1.44 g	8 mM
KH$_2$PO$_4$	0.24 g	1.46 mM

Dissolve the above in 800 ml of deionized water, adjust pH to 7.4 using HCl, and then bring the total volume to 1 l. Sterilize by autoclaving. Alternatively, many companies sell pre-made PBS, as well as PBS tablets that can be simply dissolved in water to yield the desired buffer.

LAB SESSION 15

Interim Lab

The students will use the following supplies. The following are needed per group:

- three empty snap-cap tubes
- one snap-cap tube with 3 ml of LB/kan
- sterile toothpicks
- tube racks
- Sharpies (or other marker pens)
- access to their replica plate from Lab Session 14B.

Note to prep staff:

1. Place students' empty tubes and inoculated tubes in the refrigerator until the afternoon of the day before the next lab session.
2. In the afternoon of the day before the lab session, place the 3 ml inoculated tubes in the shaking incubator at 37°C.

LAB SESSION 15

Total RNA Purification

Note to prep staff:

1. Two hours before the beginning of the lab session, induce each 3 ml culture as described in Lab Session 15B, steps 1–3.

2. Each station will need one box of barrier tips in each size: P1000, P200, P20 and P10. These tips are wrapped and certified DNase and RNase-free. They do not need to be autoclaved. These tips will be used in Lab Sessions 15, 16 and 18.

3. It is recommended that students wipe down bench tops and micropipettes with RNase*ZAP* (Ambion AM9780) immediately prior to starting each of the following labs: 15, 16 and 18.

Recipes

LACTOSE (20 mg/ml)

Dissolve 0.1 g α-lactose monohydrate in 5 ml deionized water. Filter sterilize. Store at 4°C until needed. Make fresh each time.

2× YT/KAN/IPTG

To 100 ml 2× YT, add 100 μl kanamycin (30 mg/ml stock) and 100 μl IPTG (20 mg/ml stock). *The final concentration of IPTG is 20 μg/ml or 0.08mM.*

2× YT/KAN/LAC

To 100 ml 2× YT, add 100 μl kanamycin (30 mg/ml stock) and 144 μl lactose (20 mg/ml stock). *The final concentration of lactose is 28.8 μg/ml or 0.08mM.*

LYSOZYME (40 mg/ml)

Dissolve 0.04 g lysozyme in 1 ml deionized water. This is a 100× stock. Make 10 μl aliquots, at least one aliquot per group. Each group will only require 5 μl. Store the aliquots and remaining solution at −20°C.

DNaseI (2.73 KUNITZ UNITS/I)

Add 550 μl RNase-free water provided in Qiagen DNaseI set to the vial. Cap and invert to mix. Make 35 μl aliquots. Store at −20°C and avoid multiple freeze/thaws.

The students will use the following supplies. The following is needed per group:

- 10 μl 100× lysozyme stock
- Qiagen RNeasy Plant Kit:
  - 2.3 ml Buffer RLT (containing 1% β-mercaptoethanol)
  - three Qiashredder spin columns (lilac)
  - 1.7 ml 100% ethanol
  - six RNeasy spin columns (pink)
  - 2.3 ml Buffer RW1
  - 7 ml Buffer RPE
  - 350 μl RNase-free water.
- Qiagen DNaseI set:
  - 35 μl aliquot of prepared DNaseI
  - 35 μl Buffer RDD
  - 100 μl RNase-free water.

Students will share the following:

- extra RNase-free water from Qiagen RNeasy kit for blanking the Nanodrop.

LAB SESSION 16

Analysis of *gst::egfp* mRNA Levels by RT-qPCR: Part 1

The students will use the following supplies. The following is needed per group:

- Bio-Rad iScript™ Select cDNA Synthesis Kit:
  - 18 µl random primer mix
  - 4.5 µl iScript reverse transcriptase
  - 35 µl 5× iScript select reaction mix
  - 50 µl nuclease-free water.
- three strips of PCR tubes (compatible with thermal cycler used for qPCR)
  - six tubes needed for reverse transcription, 12 tubes needed for qPCR. *We use 0.2 ml 8-tube strip without caps, natural, 125 strips (1000 tubes) (Bio-Rad TBS-0201).*
- three optical flat 8-cap strips, for 0.2 ml tubes and plates, ultraclear, 120 (Bio-Rad TCS-0803)
- 6 µl aliquot of 10 pmol/µl mix of egfp primers:
  - egfp16 F: CGACGGCAACTACAAGACC
  - egfp16 R: GTCCTCCTTGAAGTCGATGC.

Resuspend each original, lyophilized primer in sterile water to a concentration of 100 pmol/µl. Then combine 50 µl of the forward primer with 50 µl of the reverse primer, plus 400 µl water and vortex before aliquoting for students.

- 6 µl aliquot of 10 pmol/µl stock mix of 23S primers:
  - 23S 16 F: GACGGAGAAGGCTATGTTGG
  - 23S 16 R: GTTGCTTCAGCACCGTAGTG
- 195 µl Bio-Rad iQ™ SYBR® Green Supermix
- sterile water.

LAB SESSION 17

Analysis of *gst::egfp* mRNA Levels by RT-qPCR: Part 2

Analysis of RT-qPCR data will require the .odm file saved from Lab Session 16. Each station will also need access to a computer. If student computers do not have Bio-Rad MyiQ software (or other instrument-specific software), instructors can complete steps 1–17 before the lab session and have students work from a Microsoft Excel file.

LAB SESSION 18

Analysis of *gst::egfp* mRNA Levels by Semi-Quantitative RT-PCR: Part 1

The students will use the following supplies. The following is needed per group:

- Bio-Rad iScript™ Select cDNA Synthesis Kit:
  - 18 µl random primer mix

- 4.5 µl iScript reverse transcriptase
- 35 µl 5× iScript select reaction mix
- 50 µl nuclease-free water
- three strips of PCR tubes
  - 6 tubes needed for reverse transcription (part A), 12 tubes needed for PCR (part B)
  - we use 0.2 ml 8-tube strip without caps, natural, 125 strips (1000 tubes) (Bio-Rad TBS-0201).
- three flat 8-cap strips, for 0.2 ml tubes and plates, ultraclear, 120 (Bio-Rad TCS-0803)
- 4 µl of 10 pmol/µl stock mix of egfp primers:
  - egfp 18 F CCTGAAGTTCATCTGCACCA
  - egfp 18 R TGCTCAGGTAGTGGTTGTCG
- 4 µl of 10 pmol/µl stock mix of 23S primers:
  - 23S 18 F ACTGCGAATACCGGAGAATG
  - 23S 18 R CCTGTTTCCCATCGACTACG
- 40 µl 10× Taq buffer
- 8 µl dNTP mix (stock of 10 mM each dNTP)
- 3 µl Taq (5 U/µl)
- sterile water.

LAB SESSION 19

Analysis of *gst::egfp* mRNA Levels by Semi-Quantitative RT-PCR: Part 2

The students will use the following supplies. Students should have from before:

- 1 kb DNA ladder
- 10× DNA loading buffer
- GelRed (10,000× stock)
- PCR samples from Lab Session 18.

Students will share the following:

- agarose
- 1× TBE buffer.

Analysis of semi-quantitative RT-PCR data can be performed with Kodak Molecular Imaging software (or other imaging software), if available. Each station should have access to a computer. If student computers do not have imaging software, students will have to take turns at a common computer. Quantification can be performed using Microsoft Excel.

APPENDIX 3

Preparation of Competent
E. coli Cells

Introduction

Escherichia coli can be made competent for DNA uptake by treatment with divalent cations, and they can be stored in this state indefinitely at $-80°C$. Many parts of the procedure for making competent cells are critical for obtaining cells with a high transformation efficiency. The cells must be harvested in log phase, and once chilled they must not warm to room temperature or they will lose competence. For optimal transformation efficiencies, the cells should be aliquoted in a cold room.

Because the expression vector we are using, pET-41a, has a T7 promoter that drives transcription of the gene of interest, the *E. coli* host strain must be an λDE3 lysogen so that it will have a T7 RNA polymerase to bind the promoter for expression. A suitable strain is BL21(DE3).

Protocol

Preparation of Chemically Competent Cells by Calcium Chloride Treatment

Remember that you are not using antibiotics in this procedure. Use a sterile technique and make sure the centrifuge bottles are sterile.

1. Start an overnight culture of *E. coli* by using an inoculating loop to scrape cells from a single colony into 2 ml of LB in a polypropylene snap-cap tube. (Note: Antibiotics are not added.) Incubate at 37°C overnight in a shaking incubator.
2. Start 100 ml cultures of *E. coli* BL21(DE3) with a fresh overnight culture as inoculum. Based on the number of students in your class, inoculate several 1 liter flasks containing 100 ml of Luria-Bertani (LB) broth with 0.1 ml of overnight culture. You will need to start approximately one 100 ml culture per ten lab stations.
3. Incubate the flask at 37°C with shaking. Grow the bacteria to log phase. An OD_{600} value between 0.3 and 0.5 may be used. It will take 2.5–4 hours for the culture to grow to the proper stage for making competent cells.

4. Using a sterile technique and a sterile 5 ml pipette, withdraw samples from the cultures and determine the OD_{600} using a spectrophotometer. *Once cultures appear cloudy, they are in log phase growth and their OD will increase rapidly.*

5. When the proper optical density is reached, pour the culture into two pre-chilled sterile large centrifuge tubes. Balance the tubes and then place on ice for 15 minutes. From now on, it is important that the cells *never* warm to room temperature.

6. Harvest the cells by centrifugation at 6000 rpm for 10 minutes at 4°C. Make sure the rotor is pre-chilled.

7. Pour off the supernatant and, keeping the cells on ice, remove excess LB medium with a Pasteur pipette or a 1 ml micropipette. Resuspend the cells in 1 ml of ice-cold sterile transformation and storage solution (TSS).[1] Resuspend gently by pipetting up and down. Transfer the contents of one tube to the other to consolidate your cells. Be very gentle; the cells are now fragile.

   TSS contains 50 mM $MgCl_2$; other procedures use 50 mM $CaCl_2$. The dimethylsulfoxide (DMSO) in TSS is necessary if you intend to freeze some cells. Because there is little loss in transformation efficiency, it is always useful to freeze extra cells.

8. Pipette 75 μl aliquots quickly into cold microcentrifuge tubes on ice. These cells can be used immediately or frozen at −80°C. You should keep one tube on ice for the test transformation and freeze the rest. You will need one tube of frozen competent cells per station for Lab Session 5.

9. To flash freeze the competent cells, take your ice bucket to a dry/ice ethanol bath near a −80°C freezer. Using a pre-cooled microcentrifuge tube holder, submerge the tubes. Retrieve the tubes into a labeled cardboard or styrofoam box and immediately place at −80°C.

Transformation Control

The transformation efficiency is measured by determining the number of colony-forming units (cfu)/μg DNA. Only supercoiled DNA should be used (linear DNA does not transform efficiently). In this section you will use 10 ng of uncut pET-41a DNA to confirm the competence of the cells.

1. Label the tops of the two pre-chilled sterile microcentrifuge tubes with a designation for the following treatments: TE, pET-41a.

2. Aliquot 20 μl of competent cells into each tube. Remember to keep the cells on ice at all times.

3. Add to the appropriate tube:
   - 2 μl TE
   - 2 μl of a 5 ng/μl pET-41a DNA solution.

   Mix gently, and leave on ice for 5 minutes.

4. Heat pulse the cells for 30 seconds at 42°C.

5. Incubate tubes on ice for 2 minutes.

6. Add 80 μl of liquid medium (either LB or SOC), gently invert the tubes to mix and incubate at 37°C, 225 rpm for 45 minutes.

7. Plate the entire volume (100 μl) on a selective medium (LB/kan).

8. Incubate the plates at 37°C overnight.

9. Observe plates. TE control should have no colonies; pET-41a control should have numerous colonies or a lawn of growth.

Reference

1   Chung CT, Nienela SL, Miller RH. One-step preparation of competent Escherichia coli: Transformation and storage of bacterial cells in the same solution. *Proc. Nat. Acad. Sci. U.S.A.* 1989;86:2172–2175.

APPENDIX 4

Pre-Lab Questions

LAB SESSION 1

1. What is your major(s)/department (grad students – principle investigator, if chosen)?
2. What are you hoping to learn in this class?
3. What are your career goals?
4. Have you had previous experience with molecular biology/cloning? If so, what (relevant coursework/research experience)?

LAB SESSION 2

1. What are the volume ranges (max. and min. amount to be measured) of the following micropipettes?
   - P10
   - P20
   - P200
   - P1000.
2. What is the volume (in microliters) that would be pipetted using the following settings?

0		1		0
5		5		6
0		4		4
(P200)		(P20)		(P1000)

   a. _____ b. _____ c._____

3. What method will we use to separate plasmid from chromosomal DNA?
   a. Alkaline lysis
   b. Silica adsorption
   c. NanoDrop/ spectrophotometry
   d. Restriction digestion.
4. An $A_{260/280}$ of _____ indicates optimal purity of double-stranded DNA.

5. Why will you use two different restriction enzymes to cut the vector pET-41a?
   a. In case one of the enzymes fails to cut.
   b. Because the two enzymes have compatible cohesive ends.
   c. Because cutting the vector with two enzymes that leave incompatible ends and then cutting the insert with the same two enzymes will force the insert into the correct orientation when cloning.
   d. a and b.

LAB SESSION 3

1. What DNA serves as the template to amplify *egfp* in today's PCR?
   a. pET-41a
   b. egfpNco
   c. egfpNot
   d. pEGFP-N1
   e. b and c.
2. What is the basis for selecting the annealing temperature to use in a specific PCR reaction?
   a. The length and GC content of the primers.
   b. The length and GC content of the desired PCR product.
   c. The restriction sites engineered into the primers.
   d. The ideal reaction temperature of the polymerase.
3. What is the purpose of running our digested vector through the spin column?
   a. To remove any live *E. coli* that might be mixed in the sample.
   b. To remove restriction enzymes from the digest.
   c. To remove unwanted salts prior to ligation.
   d. a, b and c.
   e. b and c.
4. In DNA agarose gel electrophoresis, which side of the apparatus should your wells be closer to?
   a. The black side (cathode).
   b. The red side (anode).
5. True or False: Because a 73 base pair fragment is removed from pET-41a during digestion, the uncut vector should always run slower through the agarose gel than the cut/digested vector.

LAB SESSION 4

1. Approximately what size band do you expect to see on your PCR gel?
   a. 700 kb
   b. 5 kb
   c. 6 kb
   d. 1 kb
   e. 700 bp.
2. If your Nanodrop reading was 75 ng/ul, what volume of PCR product would you need to add to your restriction digest?
   a. 0.15 μl
   b. 6.67 μl
   c. 150 μl
   d. 6670 μl.

3. What is a purpose of taking a Nanodrop reading of your PCR product?
   a. To determine concentration of DNA.
   b. To determine the volume of DNA.
   c. To determine the fluorescence intensity of egfp
   d. To determine whether you have non-specific PCR products.
4. If your clean, digested *egfp* PCR product has a concentration of 7 ng/µl, what volume would you need to use to have 21 ng for the following week's ligation (use equation given at the end of the lab session)?
5. What should you do if you have an aliquot of enzyme and cannot pipette the entire volume out of the tube?
   a. Try to centrifuge down the liquid for 5 seconds
   b. Vortex
   c. Complain to your TA as a first resort
   d. Cry and go home.

LAB SESSION 5

1. True or False: Linear DNA will not be replicated in *E. coli* even if it is taken up into the cell.
2. True or False: Ligase buffer does not need to be kept on ice when not in use since it is active at room temperature.
3. For today's experiment, what is the vector-to-insert ratio that you will be using?
   a. 1:1
   b. 3:1
   c. 1:3
   d. 1:10.
4. What control will you include to ensure that your cells are competent?
   a. Uncut pET-41a
   b. digested vector with insert
   c. digested vector without insert
   d. none of the above.
5. True or False: Gently mix competent cells by flicking the tube instead of vortexing them before use in transformation.

LAB SESSION 6

1. True or False: IPTG is added to the membrane to induce the expression of *lacZ*, thereby allowing protein to be expressed.
2. True or False: The colony hybridization using an antibody probe experiment will tell us whether the EGFP protein is expressed by our transformants.
3. True or False: You will treat the membrane that is to be used for the colony hybridization with an antibody probe with the enzyme DNAse.
4. In the colony hybridization using an antibody probe experiment, to what molecule should the primary antibody bind?
5. What is the purpose of blocking solution?
   a. To block the primary and secondary antibodies from binding directly to the nitrocellulose membrane.
   b. To block the chloronaphthol from binding directly to the nitrocellulose membrane.

c. To block the primary antibody from binding non-specifically to *E. coli* proteins other than the GST::EGFP fusion protein.

d. a and c.

LAB SESSION 7

1. In what species is the monoclonal EGFP antibody raised?
   a. Goat
   b. jellyfish (*Aequorea victoria*)
   c. mouse
   d. rabbit.
2. Which is true about GAMP?
   a. It is the secondary antibody.
   b. It binds specifically to the α-EGFP primary antibody.
   c. It has horseradish peroxidase conjugated to one end to be used as a means of detection.
   d. All of the above.
3. A positive result on the mAb probe blot tells you that:
   a. The particular clone that appeared positive is making the protein of interest.
   b. The particular clone that appeared positive has the DNA insert of interest, but you don't know the orientation of the insert.
   c. The particular clone that appeared positive has the DNA insert of interest and it is in the correct orientation.
   d. a and c.
4. In the PCR screen, how are the primers designed?
   a. Both are designed to bind to opposite ends of the insert DNA.
   b. Both are designed to bind to vector DNA, on opposite sides of the insert.
   c. Both are designed to bind to vector DNA, on the same side of the insert.
   d. One is designed to bind insert DNA and one is designed to bind adjacent vector DNA.
5. In your PCR screen, what would you expect to see when you run your PCR product on a gel if you had a negative clone (a clone with no insert)?
   a. No band on the gel.
   b. One band about 1500 bp (1.5 kb).
   c. Two bands.

LAB SESSION 8

1. What size do you expect your PCR product to be from the screening experiment for clones that have the *egfp* gene?
2. What size do you expect your PCR product to be from the screening experiment for clones that DO NOT have the *egfp* gene?
3. True or False: The miniprep protocol you'll be performing uses alkaline lysis to purify the plasmid DNA.
4. True or False: While performing a miniprep, you need to vortex your samples well to ensure complete mixing after the addition of Buffer P2.

5. If you transferred a colony while replica plating to an LB/kan plate instead of an LB/kan/IPTG plate, what would you expect to see next week when visualizing the cells?
   a. Bacteria would not grow the plate.
   b. Bacteria would grow on the plate, but they would have not have the EGFP gene present and therefore there would be no EGFP protein expression.
   c. Bacteria containing the EGFP gene would grow, but EGFP protein would not be expressed.
   d. There would be bacteria growing with high levels of EGFP expression.

LAB SESSION 9

1. What is the purpose of preparing a master mix?
   a. Minimize error inherent in pipetting smaller volumes
   b. reduces variability between samples
   c. saves time
   d. all of the above.
2. In the double-digest restriction mapping experiment, what size fragments do you expect to see for clones that have the *egfp* gene?
3. In the double-digest restriction mapping experiment, what size fragments do you expect to see for clones that DO NOT have the *egfp* gene?
4. What precaution must you take when looking at your IPTG plate on the UV box?
5. Why does one sequence positive clones derived from PCR cloning?

LAB SESSION 10

1. Each peak in the chromatogram corresponds to:
   a. A fluorescent deoxynucleotide triphosphate (dNTP) which has been released from the DNA fragment resulting in the termination of synthesis.
   b. A fluorescent deoxynucleotide triphosphate (dNTP) which has been incorporated into the DNA fragment resulting in the termination of synthesis.
   c. A fluorescent dideoxynucleotide triphosphate (ddNTP) which has been released from the DNA fragment resulting in the termination of synthesis.
   d. A fluorescent dideoxynucleotide triphosphate (ddNTP) which has been incorporated into the DNA fragment resulting in the termination of synthesis.
2. A typical sequencing read will typically yield _____ nucleotides of unambiguous sequence.
   a. 80–100
   b. 300–500
   c. 800–1000
   d. 3000–5000.

189

3. Sequencing primers should be designed to bind where?
   a. ~60 nucleotides upstream of the beginning of the region to be sequenced.
   b. ~5 nucleotides upstream of the beginning of the region to be sequenced.
   c. ~5 nucleotides downstream of the beginning of the region to be sequenced.
   d. As long as the primer binds within the region to be sequenced, it doesn't matter exactly where it is designed.

4. Where do you expect to see multiple "Ns" within your sequencing read?
   a. In the beginning of your sequence (first 25 nucleotides)
   b. In the middle of your sequence
   c. In the end of your sequence (last 25 nucleotides)
   d. a and b
   e. a and c
   f. None of the above.

5. Which tool will you use to look for similarity between your sequencing data and the expected pBIT sequence?
   a. BLAST
   b. Sanger sequencing
   c. Chromas Lite
   d. Genbank.

LAB SESSION 11

1. In SDS-PAGE, what chemical is used to ensure that all protein molecules are coated with a negative charge?
   a. IPTG
   b. β-mercaptoethanol
   c. SDS
   d. X-gal.

2. In SDS-PAGE, what chemical is used to ensure that protein disulfide bonds are broken?
   a. IPTG
   b. β-mercaptoethanol
   c. SDS
   d. X-gal.

3. What is the advantage of running a discontinuous protein gel rather than a continuous gel?
   a. Proteins can be separated by size alone, rather than being dependent on size, shape and charge.
   b. For better resolution of the protein bands.
   c. The gel runs faster.
   d. a and b.
   e. None of the above.

4. What is the advantage of treating protein samples with SDS and β-mercaptoethanol?
   a. Proteins can be separated by size alone, rather than being dependent on size, shape and charge.
   b. For better resolution of the protein bands.
   c. The gel runs faster.

   d. a and b.
   e. None of the above.
5. The non-polymerized forms of acrylamide (powder and liquid) are far more dangerous than polymerized (solidified) polyacrylamide.
   a. True
   b. false.

LAB SESSION 12

1. What should you be able to see on your gel after staining with Ponceau Red?
   a. The molecular weight markers.
   b. Only the specific GST::EGFP fusion protein.
   c. All of the proteins expressed by *E. coli* in the lanes where you loaded cell lysates.
   d. a and b.
   e. a and c.
2. When performing the Ponceau stain, you may see a band of approximately 25 kD in your negative control that is not present in your positive control or your other positive clones. What does this band likely represent?
3. At the completion of the western blot, what will you see on the membrane?
   a. The molecular weight markers.
   b. Only the specific GST::EGFP fusion protein.
   c. All of the proteins expressed by *E. coli* in the lanes where you loaded cell lysates.
   d. a and b.
   e. a and c.
4. Approximately what size (with units) will the GST::EGFP fusion protein be?
5. Probing the western blot uses the same principles as what other experiment you have performed in this course?

LAB SESSION 13

1. What is the purpose of the GST portion of your fusion protein?
   a. It makes the protein glow green.
   b. It is responsible for the IPTG induction of the protein.
   c. It allows for the affinity purification of the protein using a glutathione affinity column.
   d. It allows the fusion protein to be cleaved into two fragments.
2. Where did the *gst* gene come from?
   a. It was already engineered into the pET-41a expression vector.
   b. Fluorescent jellyfish.
   c. We cloned it into the pET-41a expression vector after gel-purifying it.
   d. b and c.
3. Which fraction from an affinity chromatography purification experiment has the highest amount of the protein of interest?
   a. Cell lysate
   b. wash
   c. eluate.

4. Which fraction from an affinity chromatography purification experiment has the highest purity of the target protein?
   a. Cell lysate
   b. wash
   c. eluate.
5. What method(s) will you employ to lyse bacterial cells in today's lab?
   a. French Press
   b. sonication
   c. freeze–thaw
   d. exposure to chloroform vapors
   e. a and c
   f. b and c
   g. c and d.

LAB SESSION 14

1. We need to use an EGFP-specific antibody in order to assess if we have successfully purified our fusion protein from contaminants.
   a. True
   b. false.
2. Additional bands in eluate fractions, i.e. other than the band representing the GST::EGFP fusion protein, could have resulted from:
   a. Contaminants present in your sample (i.e. column not washed well).
   b. Degradation of the fusion protein.
   c. Fusion protein's inability to bind to column
   d. a and b.
   e. All of the above.
3. How will we determine the concentration of the fusion protein?
   a. Absorbance reading at 280 nm.
   b. Bradford assay.
   c. BSA assay.
   d. Fluorescence assay.
4. Following the protocol in the lab manual, you determined that the fusion protein concentration in the assay well of your fourth eluate fraction was 0.2 µg/ml. What is the concentration in the original sample? Be sure to show your units.
5. Using the concentration that you calculated in question 4, what is the total amount of fusion protein in the fourth eluate fraction if the total volume that you had collected was 0.5 ml? Show units.

LAB SESSION 15

1. List two reasons why homogenization of the E. coli is performed.
2. Why are you using a second RNeasy spin-column after the DNase step?
   a. To purify additional RNA you missed with the first column.
   b. To inactivate the Dnase.
   c. To remove salts.
   d. Both a and c.
   e. Both b and c.

3. What nucleic acid will you purify from *E. coli*?
   a. Genomic DNA.
   b. Complementary DNA (cDNA).
   c. Messenger RNA (mRNA).
   d. Total RNA.
4. What is the extinction coefficient of RNA?
5. If you read the absorbance of your samples and the A_{260}/A_{280} ratio is 1.5, what is the most likely contaminant?

LAB SESSION 16

1. What type of enzyme will be used to convert RNA to cDNA?
   a. Reverse transcriptase
   b. RNA polymerase
   c. ligase
   d. oligo-dT.
2. In qPCR, what is used to quantify how much DNA is amplified?
   a. Spectrophotometry
   b. gel electrophoresis
   c. fluorescence.
   d. luminescence.
3. Why is a no RT reaction prepared?
   a. To help quantify the level of the reference gene.
   b. To control for protein contamination.
   c. To control for DNA contamination.
   d. All of the above.
4. What are the two required qualities of a reference gene?
5. How many master mixes will you be preparing for the reverse transcription step?
   a. One
   b. two
   c. three
   d. four.

LAB SESSION 17

1. What do you call the cycle at which an amplification plot crosses the threshold?
   a. Exponential cycle
   b. threshold cycle
   c. baseline cycle
   d. plateau cycle.
2. Assuming a threshold was properly set; if amplicon A has a C_T value of 22 and amplicon B has a C_T value of 23:
   a. There was approximately twice as much amplicon B as amplicon A in the sample.
   b. There was approximately twice as much amplicon A as amplicon B in the sample.
   c. There was only slightly more amplicon B than amplicon A in the sample.

    d. There was only slightly more amplicon A than amplicon B in the sample.

3. In this lab session, what will serve as your calibrator?
   a. *egfp*
   b. *23S*
   c. no induction
   d. IPTG induction.

4. In this lab session, what will serve as your reference?
   a. *egfp*
   b. *23S*
   c. no induction
   d. IPTG induction.

5. What is one concern of setting your threshold too low?
   a. You could be analyzing amplification in the plateau phase.
   b. You could be analyzing amplification in the exponential phase.
   c. You could be analyzing background signals.
   d. Both a and b.

LAB SESSION 18

1. Which of the following are properties of both semi-quantitative PCR AND qPCR?
   a. They are "end-point" methods.
   b. They require a gel electrophoresis step.
   c. They can detect large (10-fold) changes in gene expression.
   d. a, b and c.

2. How many master mixes will you be preparing for the reverse transcription step?
   a. One
   b. two
   c. three
   d. four.

3. How will you control for the tendency of smaller amplicons to be more efficiently amplified than larger amplicons?
   a. Primer pairs for both amplicons will be mixed in the same tube.
   b. Primer lengths will all be identical.
   c. Amplicon lengths will all be identical.
   d. Extension times will all be identical.

4. How will you avoid analyzing levels in the plateau phase of amplification?
   a. The cycle numbers corresponding to the exponential phase were previously determined.
   b. Aliquots of each sample will be removed during the early cycles of the program.
   c. Reagents will be at concentrations high enough to never become limiting.
   d. The thermal cycler program will run for only 1.5 hours, ensuring the plateau is not reached.

5. Why is *23S* rRNA being amplified?
   a. It serves as a loading control.
   b. Its level is predicted to change with IPTG and/or lactose induction.

   c. Both of the above.
   d. None of the above.

LAB SESSION 19

1. What type of nucleic acid are you visualizing on the agarose gel?
    a. Genomic DNA
    b. complementary DNA
    c. messenger RNA
    d. total RNA.
2. If the "+RT" reactions worked as expected, how many bands will be visible in each lane of the gel?
    a. Zero
    b. one
    c. two
    d. three.
3. If the "−RT" reactions worked as expected, how many bands will be visible in each lane of the gel?
    a. Zero
    b. one
    c. two
    d. three.
4. In the +RT lanes, if the *23S* amplicons vary in intensity from lane to lane, what is/are the most likely reason(s)?
    a. There was a gel loading error.
    b. Varying amounts of RNA were reverse transcribed.
    c. The different inducers tested induced *23S* gene expression to varying degrees.
    d. a and b.
    e. a and c.
5. In the +RT lanes, if the *23S* amplicons do not vary in intensity from lane to lane, and the *egfp* amplicons do vary, what can you conclude?
    a. There was a gel loading error.
    b. Varying amounts of RNA were reverse transcribed.
    c. The different inducers tested induced *egfp* gene expression to varying degrees.
    d. a and b.
    e. None of the above.

Index

Printed and bound by CPI Group (UK) Ltd, Croydon, CR0 4YY

03/10/2024

01040311-0016